Recent Development of Alternating Current Field Measurement Combine with New Technology

Xin'an Yuan · Wei Li · Jianming Zhao ·
Xiaokang Yin · Xiao Li · Jianchao Zhao

Recent Development of Alternating Current Field Measurement Combine with New Technology

Xin'an Yuan
Department of Mechanical and Electrical
Engineering
China University of Petroleum (East China)
Qingdao, Shandong, China

Wei Li
Department of Mechanical and Electrical
Engineering
China University of Petroleum (East China)
Qingdao, Shandong, China

Jianming Zhao
Department of Mechanical and Electrical
Engineering
China University of Petroleum (East China)
Qingdao, Shandong, China

Xiaokang Yin
Department of Mechanical and Electrical
Engineering
China University of Petroleum (East China)
Qingdao, Shandong, China

Xiao Li
Department of Mechanical and Electrical
Engineering
China University of Petroleum (East China)
Qingdao, Shandong, China

Jianchao Zhao
Department of Mechanical and Electrical
Engineering
China University of Petroleum (East China)
Qingdao, Shandong, China

ISBN 978-981-97-4223-3   ISBN 978-981-97-4224-0 (eBook)
https://doi.org/10.1007/978-981-97-4224-0

This work was supported by Chinese University of Petroleum (East China).

This Springer imprint is published by the registered company Springer Nature Singapore Pte Ltd.
The registered company address is: 152 Beach Road, #21-01/04 Gateway East, Singapore 189721, Singapore

If disposing of this product, please recycle the paper.

Preface

With the development of alternating current field measurement (ACFM) technology, traditional single excitation frequency and single direction excitation structures cannot meet the requirements of multiple types of defect detection (such as cracks at different angles, and buried defects). New types of excitation structures and methods have been proposed, mainly including rotating electromagnetic field detection, multifrequency detection, and defects visual algorithm. First, the sensitivity of cracks detection using traditional ACFM is directional, high sensitivity for only perpendicular cracks detection, low sensitivity for other directional cracks, and even no signal for parallel cracks. To solve this problem, the rotating alternating current field measurement (RACFM) method is present which is capable of effectively giving high sensitivity detection for arbitrary underwater cracks in any directions at one pass scanning. Second, in order to solve the problem of deep defect detection, multifrequency excitation is generally used to obtain richer defect characteristic signals. The multifrequency ACFM can be used to identify surface and subsurface defects. And also, rich information can help to obtain more characteristic signals about the size of defects. Third, the defect negative problem study of ACFM generally focuses on defect quantization, two-dimensional and three-dimensional (3D) topography visual inversion of defects. The dimension of defects can be obtained by the mapping relationship between characteristic signals and defects visual signal processing algorithm. The visualization of defects in ACFM can provide more data support and visual results for structural safety assessment. The changes in the excitation structure and signal mentioned above have expanded the scope of application of ACFM and provided opportunities for the cross-integration and innovation of ACFM technology with other advanced detection methods. So this work mainly focuses on the study of the rotating alternating current field measurement (RACFM), the multifrequency ACFM, and the visualization method in ACFM, and this book can be divided into three parts. In part 1, three articles are employed to introduce the RACFM technology. In part 2, two articles are introduced to explain the multifrequency ACFM. In part 3, three articles are introduced to explain the visualization research in ACFM.

Chapter 1. "High Sensitivity Rotating Alternating Current Field Measurement for Arbitrary-Angle Underwater Cracks".

Chapter 2. "Detection of Cracks in Metallic Objects by Arbitrary Scanning Direction Using a Double U-Shaped Orthogonal ACFM Probe".

Chapter 3. "A Novel Fatigue Crack Angle Quantitative Monitoring Method Based on Rotating Alternating Current Field Measurement".

Chapter 4. "Inspection of Both Inner and Outer Cracks in Aluminum Tubes Using Double Frequency Circumferential Current Field Testing Method".

Chapter 5. "Novel Phase Reversal Feature for Inspection of Cracks Using Multi-frequency Alternating Current Field Measurement Technique".

Chapter 6. "Visual Reconstruction of Irregular Crack in Austenitic Stainless Steel Based on ACFM Technique".

Chapter 7. "Visual ACFM System Modeling and Optimization for Accurate Measurement of Underwater Cracks".

Chapter 8. "Research on High-Precision Evaluation of Crack Dimensions and Profiles Methods for Underwater Structure Based on ACFM Technique".

Qingdao, China   Xin'an Yuan
                                                                                            xinancom@163.com

                                                                                                    Wei Li
                                                                                            ronald8044@163.com

                                                                                            Jianming Zhao
                                                                                    jianmingzhao123@163.com

                                                                                            Xiaokang Yin
                                                                                    xiaokang.yin@hotmail.com

                                                                                                    Xiao Li
                                                                                            lix2020@upc.edu.cn

                                                                                            Jianchao Zhao
                                                                                            zjc_upc@163.com

Contents

High Sensitivity Rotating Alternating Current Field Measurement for Arbitrary-Angle Underwater Cracks

Abstract Alternating current field measurement (ACFM) technology has been used for sizing underwater structure cracks. However, conventional ACFM is more sensitive to cracks perpendicular to the induced current than cracks with other angles. In this paper, a rotating alternating current field measurement (RACFM) method and underwater test system are present for the detection of arbitrary-angle cracks with high sensitivity. The RACFM is proved by simulations and experiments. Arbitrary-angle cracks detection results obtained from ACFM and RACFM have shown that the RACFM method overcomes the limitation of directional detection of ACFM and effectively achieves high detection sensitivity for arbitrary-angle cracks on underwater structures.

Keywords Offshore oil & gas industry · Rotating alternating current field measurement (RACFM) · Highly sensitive inspection system · Arbitrary-angle crack detection

1 Introduction

In the past few decades, with the development in offshore oil & gas exploitation industry, the demand for key equipment, such as offshore platforms and pipelines, has increased dramatically. During the equipment's lifetime, they are suffering from a number of hazards including extreme storms, complex loads, and corrosions. Even a small crack will diminish the overall capacity of the key equipment significantly. In the past few years, several serious incidents were caused by key equipment failures in offshore oil and gas industry [1–5]. U.S. mineral management service reported that 1443 incidents occurred in offshore during 2001–2007.

According to the results of the industry and government research programs [6–8], it is important to prevent future failure of underwater structures by providing cracks information using inspection technologies in early stages. However, there are lots of challenges in underwater inspections, because the marine environment is always

X. Yuan et al., *Recent Development of Alternating Current Field Measurement Combine with New Technology*, https://doi.org/10.1007/978-981-97-4224-0_1

coming with physical, chemical, biological factors and the complex surface situations, such as attachments on underwater structures, which influence the operations and cracks inspection results [3, 9–13].

For underwater cracks inspection, visual inspection is a very useful and economical method, which relies on inspectors' ability and experience [14, 15]. However, small and narrow cracks, such as stress corrosion cracks (SCC), are not visible to the unaided eye in most cases. Magnetic particle inspection (MPI) [16, 17] is the most widely used method for underwater surface cracks inspection. MPI uses small magnetic particles, such as iron filings, to reveal and locate the surface cracks. However, its effectiveness depends on the situation of structures surface, which is similar to liquid coupled ultrasonic inspection methods [18, 19]. But high levels of surface cleaning will be costly for underwater equipment. As a non-contact inspection technology, magnetic flux leakage (MFL) [20] technology doesn't require high level surface cleaning before inspection. MFL is based on magnetizing the equipment and sensing the flux leakage. About 90% of metal loss detections for underwater pipelines are performed with MFL. But it is difficult to detect tight cracks as the flux will flow around these cracks without leakage. Eddy current testing (ECT) is widely used for the detection of surface and sub-surface flaws in conductive materials. Conventional ECT is highly sensitive to lift-off because of the variations in sensing coil's impedance [21]. In the underwater environment, the surface of structures is often uneven due to coatings and attachments. Therefore, constant lift-off is difficult to achieve, which affects the accuracy and detectability of conventional ECT.

ACFM is originally developed by University College of London (UCL) for sizing underwater cracks as an alternative to MPI, which based on the alternating current potential drop (ACPD) [22] technology. When the measurement is performed, the induced uniform alternating electromagnetic field will be disturbed by crack on structures. As shown in Fig. 1, two components of the disturbed magnetic field are measured to calculate the crack depth and length via mathematical models. The magnetic field in X direction (B_x) shows a reduction for the decrease of current density, which reflects the crack depth, while magnetic field in Z direction (B_z) shows a negative and positive peak at both end of the crack, which indicates the crack length [23, 24]. With the advantages of high tolerance to lift-off, no or little surface cleaning and accurate mathematical model, ACFM has been widely used for sizing cracks on underwater structures without calibration in the offshore oil and gas industry [25, 26].

There is a signal excitation coil driven by AC current to induce alternating current and magnetic field on metal surface in conventional ACFM technology, as shown in Fig. 2. The induced current perturbation will be the maximum when the crack is perpendicular to the induced current (known as perpendicular crack in this paper). But it will be the minimum when the crack is parallel to the induced current (known as parallel crack in this paper), as shows in Fig. 3. Due to this phenomenon, the sensitivity of conventional ACFM is directional, high sensitivity for perpendicular cracks and low sensitivity for other angle cracks and almost no signal for parallel cracks. Therefore, traditional ACFM has to scan the same area several times along

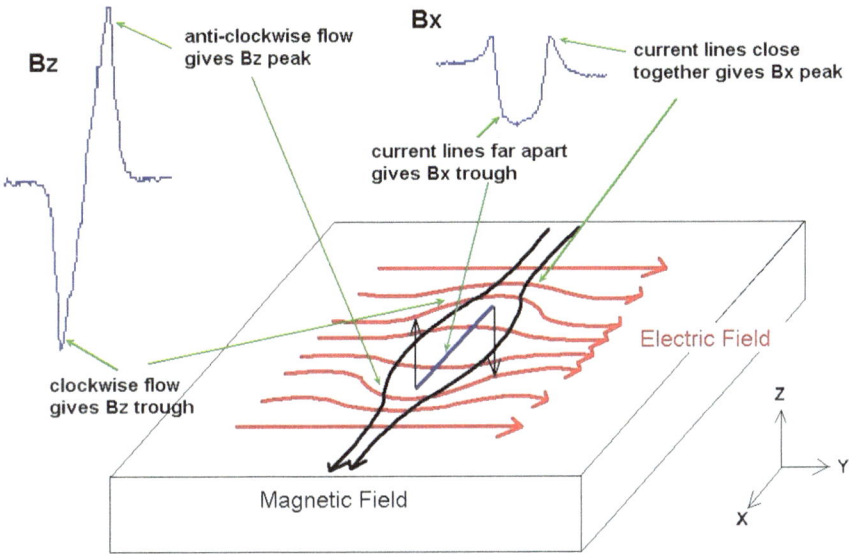

Bz

anti-clockwise flow
gives Bz peak

Bx

current lines close
together gives Bx peak

current lines far apart
gives Bx trough

Electric Field

clockwise flow
gives Bz trough

Magnetic Field

Z

Y

X

Fig. 1 Perturbations of electric field and magnetic field around a crack and ACFM signals obtained

different directions to avoid missing the cracks, which significantly increases the cost of underwater inspections. In our previous work, an optimized double U-shaped orthogonal inducer is present to detect perpendicular crack with sensor array. In this paper, a rotating alternating current field induced by the double U-shaped orthogonal inducer and an underwater test system are present for the detection of arbitrary-angle cracks on underwater structure with one pass scanning.

This paper is organized as follows: Sect. 2 shows the theoretical model of RACFM and FEM analysis of rotating alternating electromagnetic field. The RACFM system is described in Sect. 3. Arbitrary-angle underwater cracks detection experiments are conducted and discussed in Sect. 4. Conclusion is made in Sect. 5.

2 Induced Rotating Alternating Current Field

2.1 RACFM Theoretical Model

According to the ACFM principle, it is sensitive to the perpendicular crack as the distortion of induced alternating current field is most significant when the perpendicular crack presents, as shown in Fig. 3. If the induced alternating current rotates periodically, the arbitrary-angle cracks in any direction will be perpendicular to the induced field at one moment in a period, which makes it possible to have high detection sensitivity for arbitrary-angle cracks. A rotating magnetic field can be

Fig. 2 Induced AC current and Magnetic field using a signal excitation coil in traditional ACFM **a** around the excitation coil, and **b** on the metal surface

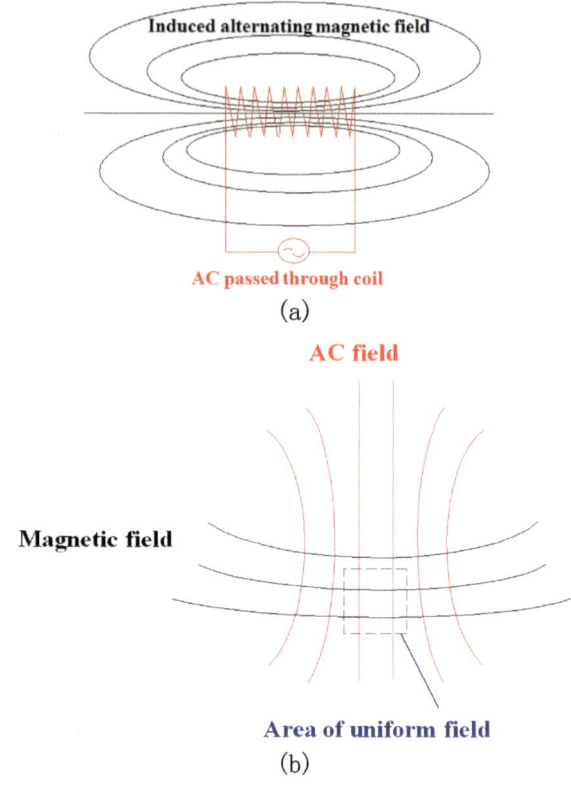

Fig. 3 Inducted AC current field perturbation caused by, **a** perpendicular crack, and **b** parallel crack

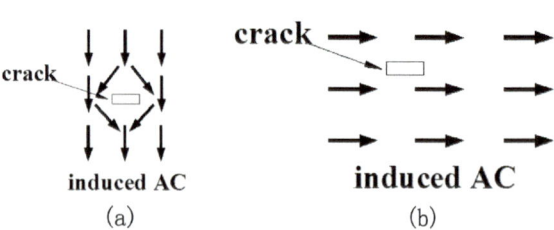

constructed using two orthogonal excitation coils with 90° phase difference alternating currents [27, 28]. In this way, two same excitation coils winding on the U-shaped MnZn ferrite yokes are placed orthogonally along X direction (excitation X) and Y direction (excitation Y) as the double U-shaped orthogonal inducer of RACFM [29], as shown in Fig. 4. Excitation X and excitation Y are driven by one pair alternating currents, $i_x(t)$ and $i_y(t)$ respectively, which are defined as follows

$$i_x(t) = I_0 \sin(\omega t + \alpha_0) \tag{1}$$

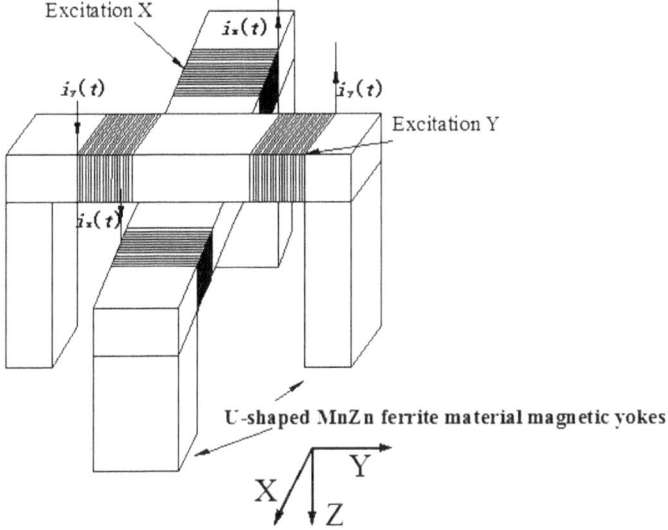

Fig. 4 Structure of double U-shaped orthogonal inducer of RACFM

$$i_y(t) = I_0 \sin(\omega t + \alpha_0 + 90°) \tag{2}$$

where I_0 is the amplitude of the alternating current, ω is the frequency of the alternating current, and α_0 is the initial phase of the $i_x(t)$. Amplitudes and frequencies of them are the same, while the initial phases are with 90° delay.

The time varying conditions of RACFM are given by Maxwell's equations as follows

$$\begin{cases} \nabla \times E = -\frac{\partial B}{\partial t} \\ \nabla \times H = J_e + \frac{\partial D}{\partial t} \\ \nabla \cdot B = 0 \\ \nabla \cdot D = \rho \end{cases} \tag{3}$$

where E is the electric field, B is the magnetic flux density, D is the electric displacement, H is the magnetic field intensity, ρ is the charge density and J_e is the current density. And the displacement current $\partial D/\partial t$ is negligible compared to the current density for the relatively low operating frequency (below 10 MHz), such as the 6k Hz frequency applied in this paper.

The constitutive relations for an isotropic, linear and homogeneous medium are given as

$$\begin{cases} B = \mu H \\ D = \varepsilon E \\ J_e = \sigma E \end{cases} \tag{4}$$

where μ is the magnetic permeability in Henrys per meter (H/m), ε is the electric permittivity in Farads per meter (F/m), and σ is the electric conductivity in Siemens per meter (S/m).

According to the Ampere's Law, when the length of excitation coil is much bigger than its radium, the alternating primary magnetic flux densities, $B_x(t)$ and $B_y(t)$, induced by the excitation X and Y, are given as follows

$$B_x(t) = \mu n i_x(t) = \mu_0 \mu_r n I_0 \sin(\omega t + \alpha_0)\vec{X} \tag{5}$$

$$B_y(t) = \mu n i_y(t) = \mu_0 \mu_r n I_0 \sin(\omega t + \alpha_0 + 90°)\vec{Y} \tag{6}$$

where μ_0 and μ_r are free space and relative magnetic permeability, n is the number of turns of each coil. $\vec{X}$ and $\vec{Y}$ just mean the directions of the primary magnetic field along X and Y directions respectively.

When the double U-shaped orthogonal inducer is closed to the conductor surface, the alternating eddy currents will be induced in the conductor. Because the double U-shaped orthogonal inducer is very close to the surface of conductor, the conductor will be assumed as a half-infinite plate. According to the principle of electromagnetic field propagation, the induced electromagnetic fields in the conductor rapidly decay exponentially with depth z, and $z = 0$ at the conductor surface. The magnetic field intensity in the conductor induced by the excitation X and excitation Y, $H_x(z, t)$ and $H_y(z, t)$, is found for time varying conditions as follows

$$H_x(z, t) = \sqrt{2}kH_p e^{-\frac{z}{d}} \cos\left(\omega t + \alpha_0 - \frac{z}{d}\right)\vec{X} \tag{7}$$

$$H_y(z, t) = \sqrt{2}kH_p e^{-\frac{z}{d}} \cos\left(\omega t + \alpha_0 + 90° - \frac{z}{d}\right)\vec{Y} \tag{8}$$

$$d = \sqrt{2/\omega\sigma\mu} \tag{9}$$

where d is the skin depth, H_p is the amplitude of primary magnetic field intensity and k is the ratio of the magnetic field on the surface of conductor to the total primary field.

As shown in Fig. 5, combining the magnetic field intensity with Maxwell's equations, the current densities induced by the excitation X and excitation Y, $J_{ex}(z, t)$ and $J_{ey}(z, t)$ [30], are given as follows.

$$J_{ex}(z, t) = \frac{2kH_p}{d} e^{-\frac{z}{d}} \cos\left(\omega t + \alpha_0 - \frac{z}{d} + \frac{\pi}{4}\right)\vec{Y} \tag{10}$$

$$J_{ey}(z, t) = \frac{2kH_p}{d} e^{-\frac{z}{d}} \cos\left(\omega t + \alpha_0 - \frac{z}{d} + \frac{3\pi}{4}\right)\vec{X} \tag{11}$$

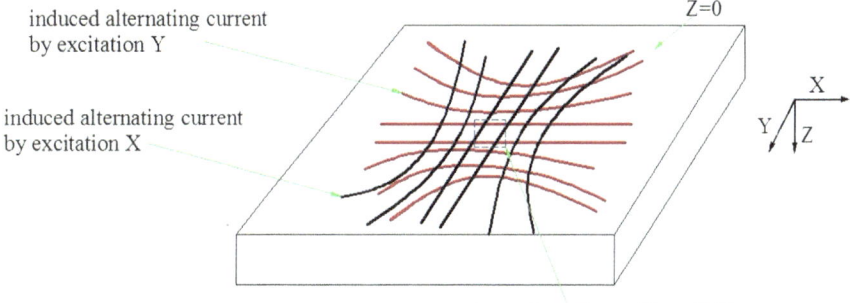

induced alternating current
by excitation Y

induced alternating current
by excitation X

Z=0

Area of rotation uniform field

Fig. 5 Induced alternating current on the surface of ample by excitation X and excitation Y respectively

The induced alternating current density, $J_e(z, t)$, can be combined with these two orthogonal components, $J_{ex}(z, t)$ and $J_{ey}(z, t)$. In this way, the amplitude and phase of $J_e(z, t)$, $A_J(t)$ and $\theta_J(t)$, are calculated as follows

$$A_J(z) = \sqrt{J_{ex}(z, t)^2 + J_{ey}(z, t)^2} = \frac{2kH_p}{d}e^{-\frac{z}{d}} \tag{12}$$

$$\theta_J(z) = \arctan\left(\frac{J_{ex}(z, t)}{J_{ey}(z, t)}\right) = \omega t + \alpha_0 - \frac{z}{d} + \frac{3\pi}{4} \tag{13}$$

According to Eqs. (12) and (13), the combined induced alternating current in the conductor also decays exponentially with depth. At a given depth, the amplitude of the induced current density is constant, while the phase is rotating periodically at the same frequency with driving alternating current of double U-shaped orthogonal inducer, as shown in Fig. 6.

According to the theoretical model, the rotating alternating current field is induced in the conductor using the double U-shaped orthogonal inducer and the set of driving currents have been presented in this paper. Thus, there is no directional limitation for crack detection using RACFM.

Fig. 6 The theory analysis results for induced alternating current field in the conductor, and T is the period

Combined induced alternating current filed

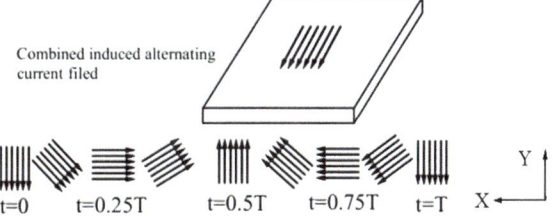

t=0   t=0.25T   t=0.5T   t=0.75T   t=T

2.2 FEM Modeling and Analyzing

To verify the RACFM theoretical model proposed above, a FEM model of the double U-shaped orthogonal inducer is set up and analyzed by using transient analysis method in ANSYS. The simulation model consists of a double U-shaped orthogonal inducer wound with excitation coils above a mild steel sample, as shown in Fig. 7. Each coil is wound by 500 turns 0.15 mm enameled copper wire. Other structural parameters of the model are given in Table 1. To simulate the real underwater environment, the computational domain of the FEM model is set to sea water. The material characteristic parameters of both model and environment are shown in Table 2 [31, 32]. The excitation coils X and Y carry the alternating currents with 1V amplitude, 6 k Hz frequency, and the 0° and 90° initial phases respectively.

A complete period is divided into 4 transient steps equally. And the transient induced current densities on the surface is simulated and analyzed. Simulation results show that the direction of induced current field at the uniform area revolves periodically at 6k Hz frequency, whose direction is negative Y at t = 0, and negative X at t = 0.25 T, and positive Y at t = 0.5 T, and positive X at t = 0.75 T, as shown in Fig. 8.

The average amplitudes of induced magnetic flux densities are almost constant at the approximate uniform area on the surface at different transient times, as shown in Table 3. The biggest relative change of average amplitudes is only 4.27%. Comparing Fig. 6 with Fig. 8, it can be seen that the uniform rotating alternating current field is induced by the double U-shaped orthogonal inducer with constant amplitude

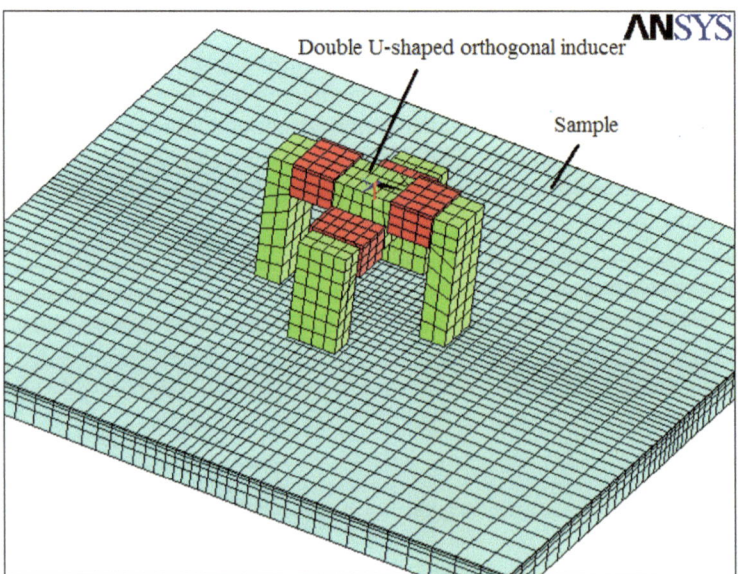

Fig. 7 The FEM model of RACFM

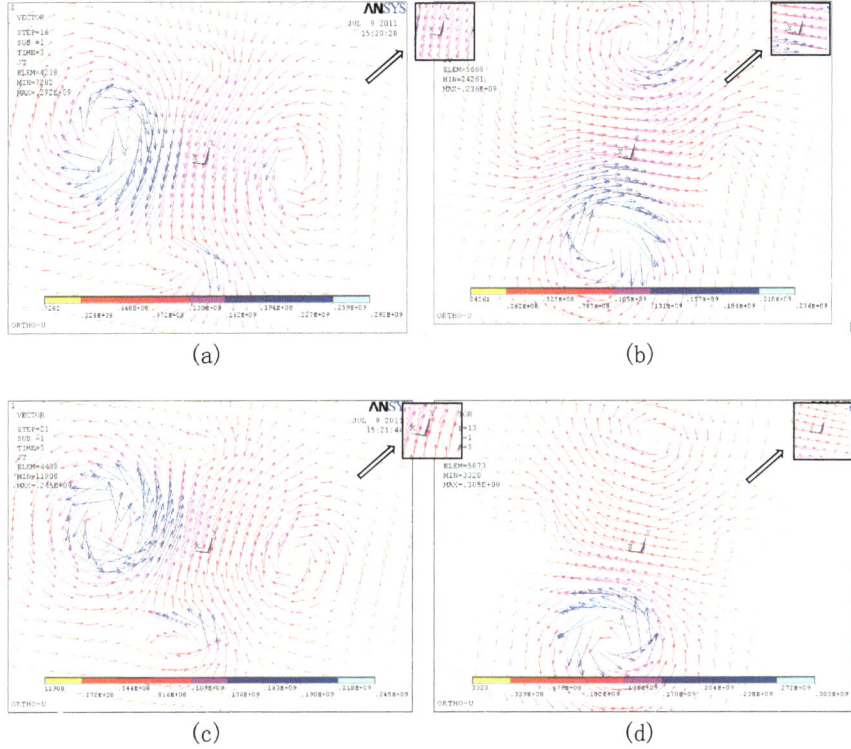

Fig. 8 The FEM simulation analysis results for induced AC field on the surface at different transient time, **a** t = 0, **b** t = 0.25T, **c** t = 0.5T, and **d** t = 0.75T

and periodically revolving direction, which matches the results of RACFM theory analysis.

Summarizing the results from theoretical analysis and FEM simulation, the uniform rotating alternating current field induced by the double U-shaped orthogonal inducer meets the requirement of overcoming the limitation of directional inspection by conventional ACFM and achieves high sensitive detection for arbitrary-angle cracks with one pass scanning.

3   RACFM System for Arbitrary-Angle Cracks Measurement

3.1   RACFM System

The underwater RACFM system consists of two parts, underwater component and topside component, as shown in Fig. 9. The topside component includes power supply, data acquisition (DAQ) module and PC. The underwater component consists of signal processing hardware including signal generator, power amplifier, 90° phase shifter and condition circuit. The RACFM probe includes double U-shaped orthogonal inducer and detecting sensor.

The power supply provides 12V DC for the signal processing hardware. The signal generator provides a sine exciting signal (6k Hz and 1 V) as the initial driving signal. The initial driving signal and the orthogonal driving signal provided by the 90°phase shifter are used to drive the excitation coils of double U-shaped orthogonal inducer. The RACFM probe scans the surface of sample and the detecting sensors pick up the distorted magnetic field above the cracks. After signal amplification and low pass filtering, these analog signals are transformed to digital signals and sent into the PC using the DAQ module. The real-time curves of B_x and B_z are plotted and the cracks can be determined.

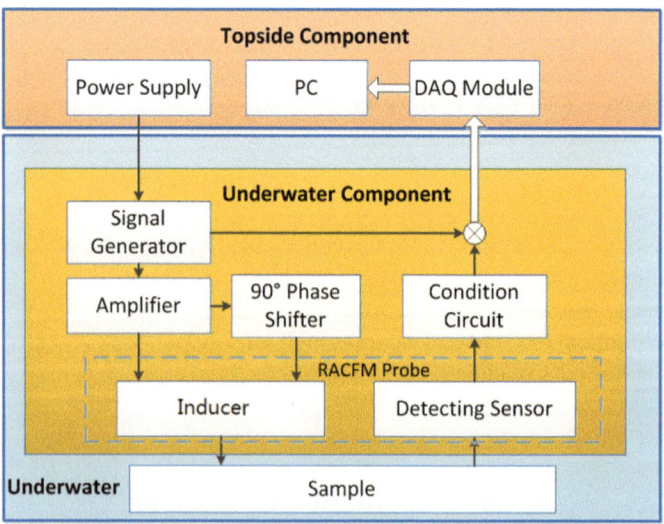

Fig. 9 The underwater RACFM system

3.2   RACFM Probe

The underwater RACFM probe consists of the double U-shaped orthogonal inducer and detecting sensors, as shown in Fig. 10a. According to theoretical model and FEM model, the double U-shaped orthogonal inducer is built by two orthogonal U-shaped MnZn ferrite yokes wound with 500 turns excitation coils of 0.15 mm enameled copper wire respectively. As shown in Fig. 10b, the detecting sensors are made up of two detection coils (B_x coils and B_z coils), which are wound on a common yoke. The planes of the two detection coils are perpendicular to X direction and Z direction for picking up the B_x and B_z respectively.

3.3   RACFM Waterproof Shell

The RACFM probe and the signal processing hardware are fixed together and encapsulated in a waterproof shell, as shown in Figs. 11 and 12. The shell material is non-magnetic stainless steel (00Cr17Ni14Mo2) and the bottom of the shell is made of non-magnetic Perspex. The detecting sensor is installed on the cover of probe at a 4 mm lift-off above the specimen. To seal against the water pressure, the gland and the sealing ring are used to compress the bottom cover together with the shell and all the signal wires pass through the cable sealing joint. The signals are transmitted via signal wires between the underwater component and topside component.

4   RACFM System Testing and Discussing

4.1   Experiment System

To further verify the effectiveness of RACFM method and underwater test system present in this paper, arbitrary-angle cracks on underwater sample detection experiments are conducted with a high precise 3D scanner, as shown in Fig. 13. A water tank filled with seawater is used to simulate the seawater environment. The RACFM underwater components and the sample are placed in water tank as in the FEM simulation. The sample under test (SUT) is a Q235 mild steel sheet with a 45mm length and 7mm depth artificial rectangular crack produced by the electric discharge machining (EDM).

   To simulate the arbitrary-angle cracks detection, the probe scans the crack with 10 different paths accurately controlled by the scanner. The angles between the scanning paths and crack change from 0° to 90° with 10° step, as shown in Fig. 14. The 0° angle crack indicates that the scanning path is along the crack, while the 90° angle crack indicates that the scanning path is perpendicular to the crack.

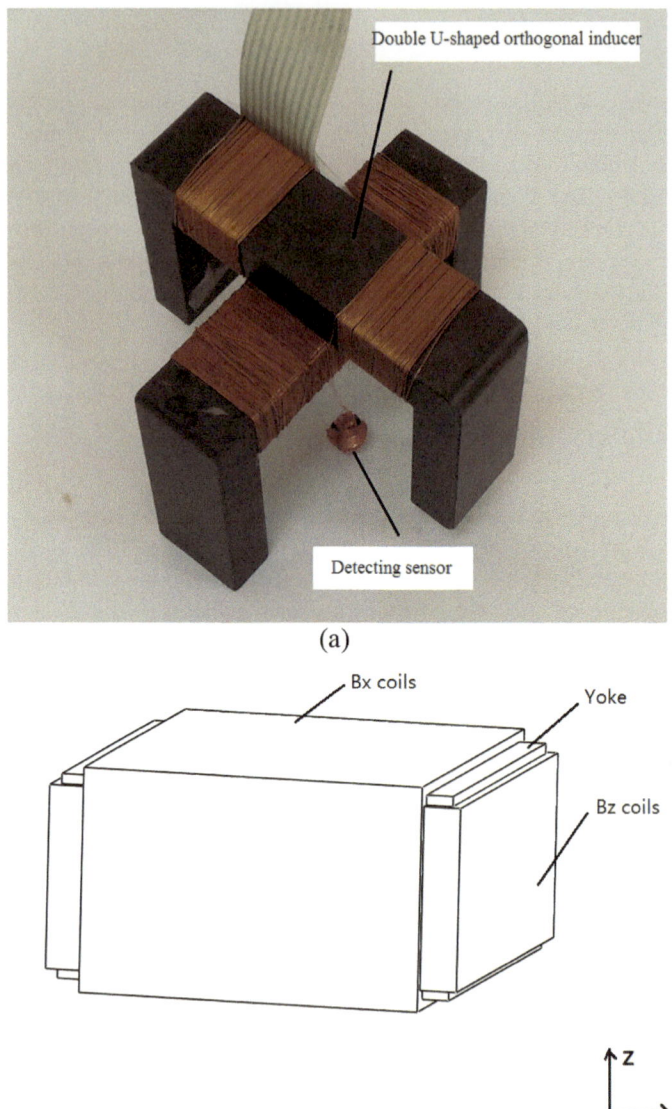

(a)

(b)

Fig. 10 Photo of RACFM probe

Fig. 11 Photo of the RACFM underwater component

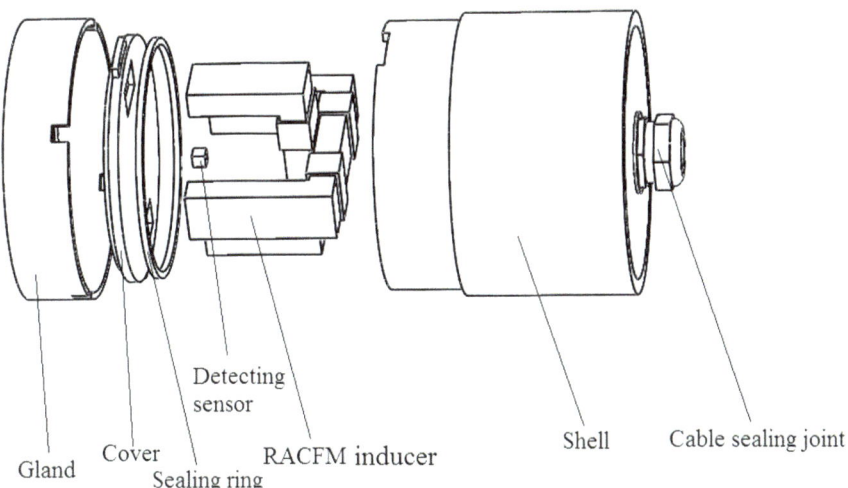

Detecting sensor

Cover

Gland

Sealing ring

RACFM inducer

Shell Cable sealing joint

Fig. 12 Assembling of RACFM underwater component

4.2 Discussion

In order to compare the detection sensitivity of different angle cracks using RACFM and conventional ACFM [33, 34], the same crack on underwater structure detection experiments have been conducted along different scanning paths using the RACFM probe and a U-shaped ACFM probe respectively. Furthermore, to keep the electromagnetic field signals picked up by different probes comparable, the excitation

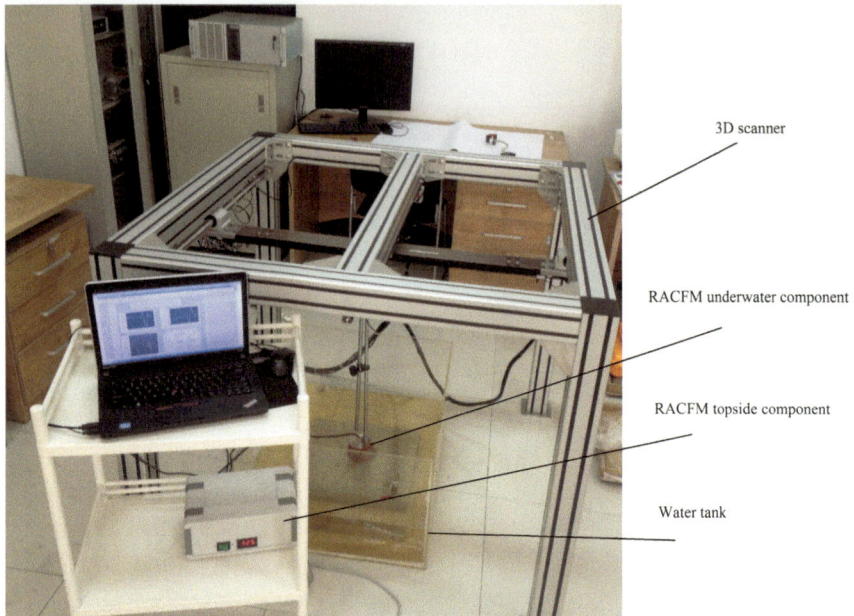

Fig. 13 RACFM experiment system

signal of the ACFM probe are the same as the excitation X of the RACFM probe. Meanwhile the detecting sensors of these two probes are the same.

When the probes scanning over the surface of sample, two components of the disturbance magnetic field, B_x and B_z, are picked up and drawn in real–time by the software after both hardware signal processing and digital signal processing. Figure 15a–d show the 0, 30, 60, 90 degree angle cracks detection experimental results using the RACFM probe and traditional ACFM probe respectively. Comparing Fig. 15a with Fig. 1, it is clear that the B_x and B_z signals obtained from experiments are in accordance with the principle of ACFM, which proves the capability of RACFM system for detection of cracks on underwater structure [34].

Comparing these experimental results of RACFM with traditional ACFM at 0 degree angle crack detection, as shown in Fig. 15a, it is apparent that the B_x and B_z distributions agree with the principle of ACFM. Meanwhile, the perturbations of B_x and B_z caused by the crack are similar using both ACFM and RACFM. When the angle is increased to 30°, as shown in Fig. 15b, the perturbations of ACFM experimental results are far smaller than those of RACFM, although both the B_x and B_z distributions can also describe the presence of the crack using both RACFM and ACFM. When the angle is increased to 60°, as shown in Fig. 15c, the B_x and B_z distributions of RACFM experimental results are still in accord with the principle of ACFM and the perturbations are big enough for recognizing the crack. However, the B_x and B_z perturbations of ACFM experimental results are small and almost covered by noise, which cannot be used to determine the crack. Finally, when the

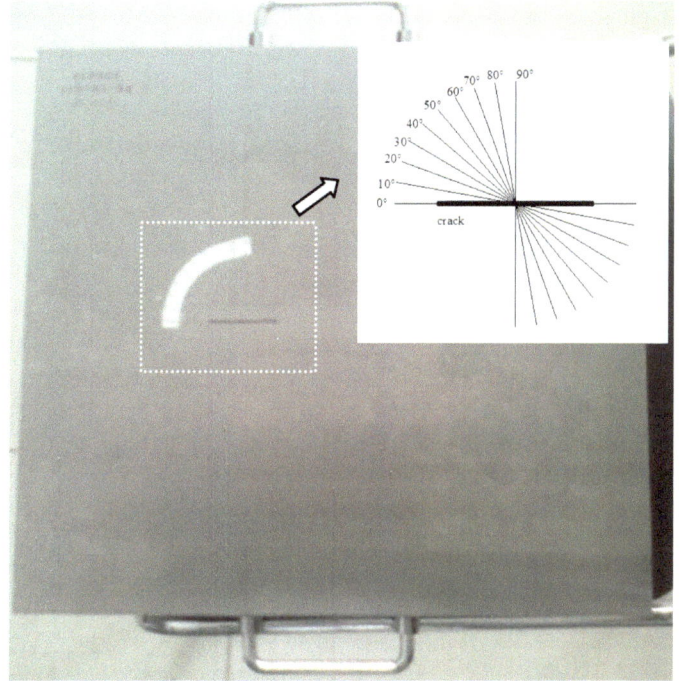

Fig. 14 Experiment sample and scanning path

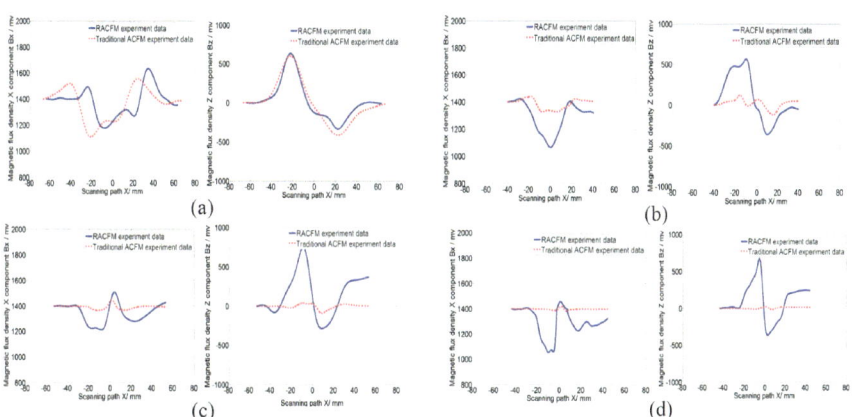

Fig. 15 B_X and B_Z signals from experiments using the RACFM probe and traditional ACFM probe for, **a** 0 degree crack detection, **b** 30 degree crack detection, **c** 60 degree crack detection, **d** 90 degree crack detection

scanning path is perpendicular to the crack (90°), as shown in Fig. 15d, the crack can be recognized by B_x and B_z distributions of RACFM experiments results perfectly, while there are no perturbations caused by the crack in ACFM experimental results.

According to the principle of ACFM, the perturbations of B_x and B_z are analyzed to determine and recognize the crack. The detection sensitivities (S_x and S_z) are given as follows to describe the maximum distorted signals above the crack [32].

$$\begin{cases} S_x = \frac{MX_{max}}{MX_0} \times 100\% \\ S_z = \frac{MZ_{max}}{MX_0} \times 100\% \end{cases} \tag{14}$$

where MX_0 is the amplitude of B_x signal without crack, and MX_{max} and MZ_{max} are the maximum perturbations of B_x and B_z caused by crack respectively.

Table 4 shows the sensitive parameters (S_x and S_z) from the experimental results of RACFM and traditional ACFM. When detecting cracks using traditional ACFM probe, the fixed direction induced electromagnetic field can be mostly disturbed by 0 degree angle crack, and the sensitive parameters decrease sharply with the increasing of the cracks' angle as shown in Fig. 16a. When the angle reaches 40 degree, the S_x and S_z are only 2.9% and 6.6%, which are less than 10% of the maximum sensitivity. Compared with traditional ACFM experimental results, the rotating alternating current field induced by the double U-shaped orthogonal inducer is disturbed by arbitrary-angle cracks. There is a little decrease of detection sensitivity for different angle cracks, as shown in Fig. 16b. The minimum sensitivities of S_x and S_z for detecting different angle cracks are still as high as 27.3% and 59.7%, which are almost 80% of the maximum sensitivity. It is apparent that the RACFM system proposed in this paper can solve the problem of directional detection and achieve high detection sensitivity for arbitrary-angle underwater cracks.

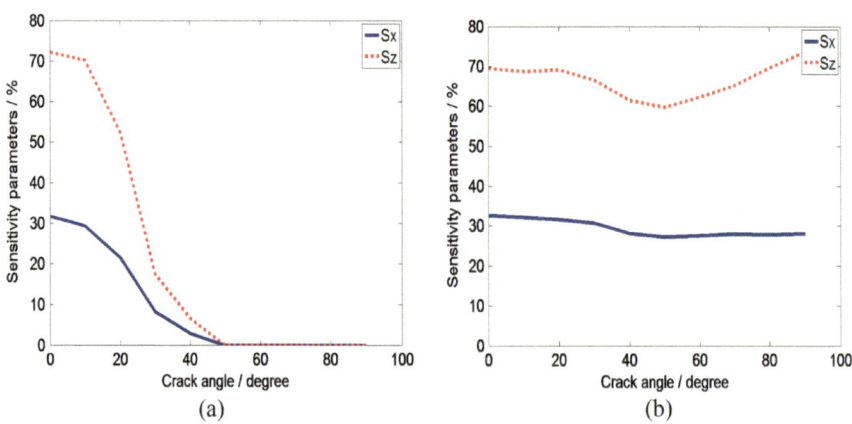

(a)   (b)

Fig. 16 Inspection sensitivity for different angle of cracks using, **a** traditional ACFM probe, and **b** RACFM probe

Table 1 The structural parameters of the FEM model

Model	Length X (mm)	Width Y (mm)	Depth Z (mm)
Yoke	90	15	36
Sample	400	400	20

Table 2 Material Parameters of the FEM model

Model	Material	Relative permeability	Resistivity ($\Omega \cdot$m)
Sample	Q235 mile steel	210	1.43×10^{-7}
Yoke	MnZn ferrite	5500	1.5×10^{4}
Coil	400	400	20
Environment	Sea water	1	2.0×10^{-1}

Table 3 The FEM simulation result at different transit time

FEM transient (T^a)	Magnetic flux density (G^b)
0	30.75
0.25	29.20
0.5	30.37
0.75	30.80

[a] $T =$ period
[b] Magnetic flux density unit G = Gauss

5 Conclusion

In this paper, the RACFM technique based on the induced rotating alternating current field is proposed to solve the directional detection problem of traditional ACFM. The rotating alternating current field is proved by the FEM simulations. Based on theoretical and simulation results, the underwater RACFM test system and encapsulated probe are present for the detection of arbitrary-angle cracks. The performance and efficiency of the RACFM system is clearly demonstrated by arbitrary-angle cracks detection experiments.

Comparing the experimental results of RACFM system with traditional ACFM system, the detection sensitivity of RACFM system does not decay too much for the non-perpendicular crack. It is clear that the RACFM overcomes the limitation of directional detection of traditional ACFM. Comparing with traditional ACFM, the RACFM system almost achieves a relatively high sensitivity for the detection of arbitrary-angle cracks on underwater structure with one pass scanning. The RACFM system can help to prevent future failures of key equipment during their whole lifetime and keep the safety of offshore oil & gas exploitation systems.

Table 4 The sensitivity of RACFM and ACFM experiments

Crack angle/degree	ACFM results		RACFM results	
	Sx/(%)	Sz/(%)	Sx/(%)	Sz/(%)
0	31.7	72.2	32.6	69.5
10	29.3	70.2	32.2	68.6
20	21.6	52.4	31.6	69.1
30	8.2	17.4	30.7	66.5
40	2.9	6.6	28.1	61.5
50	–	–	27.3	59.7
60	–	–	27.6	62.3
70	–	–	28.0	65.2
80	–	–	27.8	69.5
90	–	–	28.1	73.8

References

1. E. Nyman, Offshore oil development and maritime conflict in the 20th century: a statistical analysis of international trends. ERSS **6**, 1–7 (2015)
2. H. Fang, M. Duan, Special problems of deep-sea oil and gas engineering, in *Offshore Operation Facilities* (Gulf Professional Publishing, Boston, 2014), pp. 537–686
3. Z. Xu, M. Zi, H. Ying, H. Jin, A novel robot system for surface inspection and diameter measurement of large size pipes, in *Proceedings of 6th IEEE International Conference on Industrial Informatics* (2008), pp. 1717–1721
4. F. Carvalho, A. Raposo, I. Santos, M. Galassi, Virtual reality techniques for planning the offshore robotizing, in *Proceedings of 12th IEEE International Conference on Industrial Informatics* (2014), pp. 353–358
5. Assault on America: a decade of petroleum company disaster, pollution, and profit (National Wildlife Federation, 2010)
6. Technology strategy for Deepwater and Subsea Production Systems 2008 update (TTA group companies and organisation, 2008)
7. P.W. Sammarco, S.R. Kolian, R.A.F. Warby, J.L. Bouldin, W.A. Subra, S.A. Porter, Distribution and concentrations of petroleum hydrocarbons associated with the BP/deepwater horizon oil spill, gulf of mexico. Mar. Pollut. Bull.Pollut. Bull. **73**, 129–143 (2013)
8. P.A.S. Mendes, J. Hall, S. Matos, B. Silvestre, Reforming Brazil's offshore oil and gas safety regulatory framework: lessons from Norway, the United Kingdom and the United States. Energ Policy **74**, 443–453 (2014)
9. R.M.C. Ratnayake, Challenges in inspection planning for maintenance of static mechanical equipment on ageing oil and gas production plants: the state of the art, in *OMAE*, Rio de Janeiro (2012), pp. 91–103
10. L. Muehlenbachs, M.A. Cohen, T. Gerarden, The impact of water depth on safety and environmental performance in offshore oil and gas production. Energy Policy **55**, 699–705 (2013)
11. P. Rizzo, Sensing solutions for assessing and monitoring underwater systems. Sens. Technol. Civ. Infrastruct. **56**, 525–549 (2014)
12. B. Xiaolong, F. Yuming, L. Weisi, W. Lipo, J. Bing-Feng, Saliency-based defect detection in industrial images by using phase spectrum. IEEE Trans. Ind. Inform. **10**, 2135–2145 (2014)
13. T. Du-Ming, I.Y. Chiang, T. Ya-Hui, A shift-tolerant dissimilarity measure for surface defect detection. IEEE Trans. Ind. Inform. **8**, 128–137 (2012)

14. Y.K. Zhu, G.Y. Tian, R.S. Lu, H. Zhang, A review of optical NDT technologies. Sensors-Basel **11**, 7773–7798 (2011)
15. D. Naso, B. Turchiano, P. Pantaleo, A fuzzy-logic based optical sensor for online weld defect-detection. IEEE Trans. Ind. Inform. **1**, 259–273 (2005)
16. R. Mohammad, A.R. Mohammad, F. Hamed, Heat residual stress measurement of welded areas in steel pipes via magnetic particle testing. Mater. Eval.Eval. **70**, 624–630 (2012)
17. G. Huijun, D. Changxing, S. Chunwei, M. Jiangyuan, Automated inspection of E-shaped magnetic core elements using K-tSL-center clustering and active shape models. IEEE Trans. Ind. Inform. **9**, 1782–1789 (2013)
18. F. Honarvar, F. Salehi, V. Safavi, A. Mokhtari, A.N. Sinclair, Ultrasonic monitoring of erosion/corrosion thinning rates in industrial piping systems. Ultrasonics **53**, 1251–1258 (2013)
19. S. Hata, J. Hayashi, I. Ishimaru, S. Morimoto, Nano-level 3-D measurement system using 3-wavelength laser light interference, in *Proceedings of 6th IEEE International Conference on Industrial Informatics* (2008), pp. 721–725
20. M.M. Tehranchi, S.M. Hamidi, H. Eftekhari, M. Karbaschi, M. Ranjbaran, The inspection of magnetic flux leakage from metal surface cracks by magneto-optical sensors. Sens. Actuators, A **172**, 365–368 (2011)
21. H. Wang, Z. Feng,Ultrastable and highly sensitive eddy current displacement sensor using self-temperature compensation. Sens. Act. A Phys. **203**, 362–368 (2013)
22. W. Li, X.A. Yuan, G. Chen, X. Yin, J.H. Ge, A feed-through ACFM probe with sensor array for pipe string cracks inspection. NDT & E Int. **67**, 17–23 (2014)
23. W.D. Dover, R. Collins, D.H. Michael, The use of AC-field measurements for crack detection and sizing in air and underwater. Phil. Trans. R. Soc. Lond. A **320**, 271–283 (1986)
24. J. Zhou, W.D. Dover, Electromagnetic induction in anisotropic half-space and electromagnetic stress model. J. Appl. Phys. **83**, 1694–1701 (1998)
25. M.C. LUGG, The first 20 years of the A.C. field measurement technique, in WCNDT, South Africa (2012), pp. 16–20
26. G.M. Chen, W. Li, Z.X. Wang, Structural optimazation of 2-D array probe for alternating current field measurement. NDT&E Int. **40**, 455–461 (2007)
27. O. Yamashita, Effect of the surface integral in the torque equation of the electromagnetic angular momentum on the Faraday rotation. Opt. Commun.Commun. **284**, 4248–4253 (2011)
28. H. Hoshikawa, K. Koyama, Basic study of a new ECT Probe using uniform rotating direction eddy current, in *Review of Progress in Quantitative Nondestructive Evaluation* (1997), pp. 1067–1074
29. T.E. Capobianco, Rotating field eddy current probe for characterization of cracking in non-magnetic tubing, in *Review of Progress in Quantitative Nondestructive Evaluation* (1999; 18A–18B), pp. 449–454
30. R.F. Harrington, Some theorems and concepts, in *Time-Harmonic Electromagnetic Fields*, 2st edn. (IEEE Press, 2001), pp. 95–132
31. W. Li, G.M. Chen, C.R. Zhang, T. Liu, Simulation analysis and experimental study of defect detection underwater by ACFM probe. China Ocean Eng. **27**, 277–282 (2013)
32. L. Talley, G. Pickard, W. Emery, J. Swift, Physical properties of seawater, in *Descriptive Physical Oceanography* (2011), pp. 29–65
33. W. Li, G.M. Chen, X.K. Yin, C.R. Zhang, T. Liu, Analysis of the lift-off effect of a U-shaped ACFM system. NDT&E Int. **53**, 31–35 (2013)
34. W. Li, G.M. Chen, W.Y. Li, Z. Li, F. Liu, Analysis of the excitation frequency of a U-shaped ACFM system. NDT&E Int. **44**, 324–328 (2011)

Detection of Cracks in Metallic Objects by Arbitrary Scanning Direction Using a Double U-Shaped Orthogonal ACFM Probe

Abstract Alternating current field measurement (ACFM) technology is developed for sizing cracks on structures. According to the theory of ACFM, the induced current should be perpendicular to the crack. So traditional ACFM probe needs to scan along the cracks. In this paper, a double u-shaped orthogonal ACFM probe is present for cracks detection by induced rotating current field at arbitrary scanning direction on structures. The finite element method (FEM) model is employed to analyze the rotating current field induced by the double u-shaped orthogonal ACFM probe. The cracks detection experiments are carried out to test cracks at arbitrary scanning direction. Results show that the crack can be detected effectively at arbitrary scanning direction using the double u-shaped orthogonal ACFM probe.

Keywords ACFM · Arbitrary scanning direction · Double u-shaped orthogonal probe

1 Introduction

Alternating current field measurement (ACFM) is originally developed for sizing cracks on structures as an alternative to MPI [1, 2], based on the alternating current potential drop (ACPD) technology [3, 4]. ACFM probe induces a uniform alternating current on the surface of conductive specimen. The uniform current will be disturbed when crack presents. The distortion of magnetic field caused by electric field can be measured to deduce both the crack depth and length. As shown in Fig. 1, B_X (X direction) shows a deep trough, which contains depth information, while B_Z (Z direction) shows a negative and positive peak at both end of the crack, which gives an indication of length [5, 6]. However, the theory of ACFM is based on the assumption that the induced current is perpendicular to the crack. As shown in Fig. 2a, when the u-shaped ACFM probe scans the specimen along the crack, the induced current will be perpendicular to crack. Thus, the perturbation of induced current is obvious and the characteristic signals are in accord with the theory of ACFM [7]. While the u-shaped ACFM probe is perpendicular to the crack, as shown in Fig. 2b, the induced

© The Author(s) 2024

X. Yuan et al., *Recent Development of Alternating Current Field Measurement Combine with New Technology*, https://doi.org/10.1007/978-981-97-4224-0_2

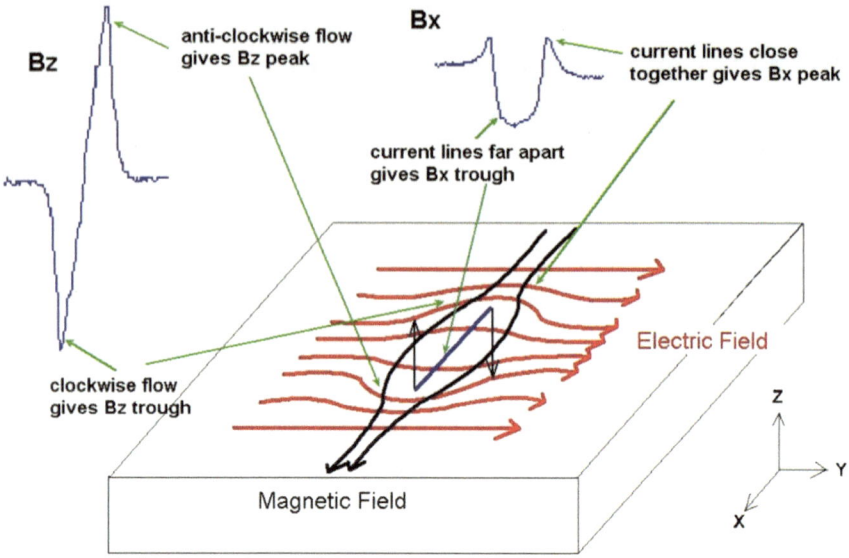

Fig. 1 Theory of ACFM

current will be parallel to the crack. In this case, the perturbation of induced current is unconspicuous. Due to leakage flux effect, some magnetic field leaks into the air. However, the leakage flux effect does not match the theory of ACFM, which can not make some quantitative detection of the crack (depth and length).

As mentioned above, the traditional ACFM probe needs to scan the specimen along the crack. In practice, it is not clear that the crack is present or not on the specimen, let alone the direction of the crack before the inspection. Hence, to detect the cracks at arbitrary scanning direction effectively, a double u-shaped orthogonal ACFM probe is present in this paper. The double u-shaped orthogonal ACFM probe can produce a rotating alternating current field on the specimen. The induced rotating current field will be perpendicular to the crack at any scanning direction. Thus the cracks can be detected at arbitrary scanning direction using the double u-shaped orthogonal ACFM probe.

2   FEM Model of Double U-Shaped Orthogonal ACFM Probe

As Ferraris effect, a rotating magnetic field can be constructed using two orthogonal coils with 90° phase difference in their AC currents [8]. In this way, two same excitation coils winding on the U-shaped MnZn ferrite material magnetic yokes are placed orthogonally along X direction (excitation X) and Y direction (excitation Y). Excitation X and excitation Y are driven by one pair alternating currents, $i_x(t)$ and

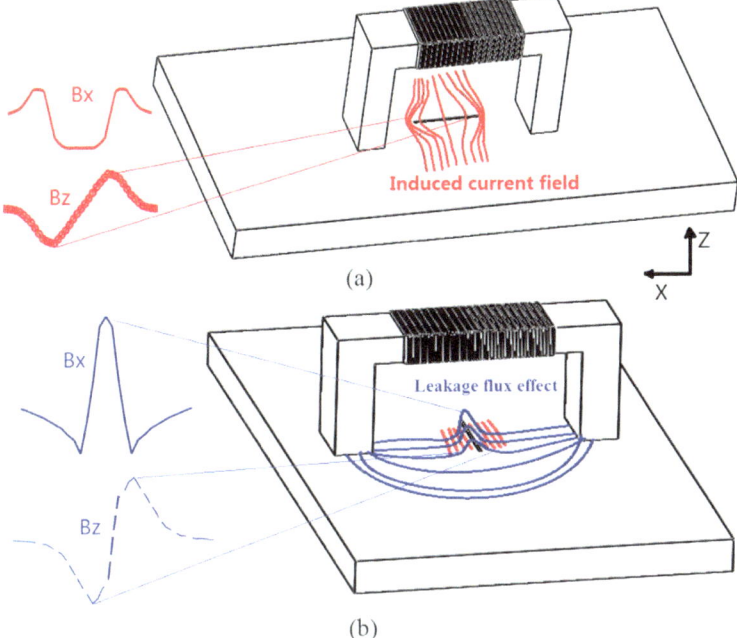

Fig. 2 U-shaped ACFM probe for cracks detection. **a** The probe is parallel to the crack. **b** The probe is perpendicular to the crack

$i_y(t)$ respectively, which are defined as follows:

$$i_x(t) = I_0 \sin(\omega t + \alpha_0) \tag{1}$$

$$i_y(t) = I_0 \sin(\omega t + \alpha_0 + 90°) \tag{2}$$

where, I_0 is the amplitude of the alternating current, ω is the frequency of the alternating current, and α_0 is the initial phase of the $i_x(t)$.

Because the excitation array is very close to the conductor surface, the conductor will be assumed a half-infinite plate [9, 10]. According to the principle of electromagnetic field propagation, the alternating eddy currents will be induced in the conductor by the alternating primary magnetic fields [11].

The FEM model of the orthogonal excitation array is set up and analyzed by the transient analysis method in ANSYS [12]. The simulation model consists of a double orthogonal U-shaped yoke wound with coils where placed on a mild steel sheet sample, as shown in Fig. 3. The excitation coils X and Y carry the alternating currents with 1 V amplitude, 6000 Hz frequency, and the initial phases of 0° and 90° respectively.

A complete period is divided into 4 transient steps equally. And the transient induced current densities on the surface are simulated and analyzed. The simulation

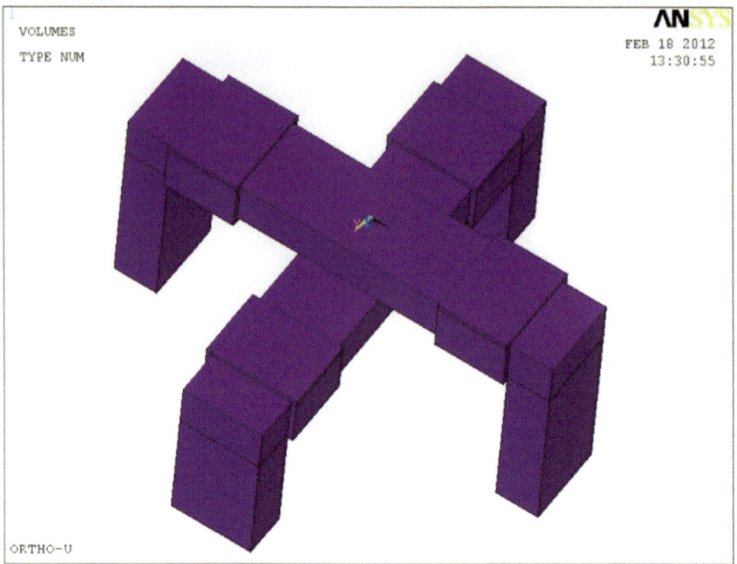

Fig. 3 The FEM model of the double u-shaped orthogonal ACFM probe

results show that the direction of induced current field at the uniform area revolves periodically with the driving alternating current, which the direction is negative Y at $t = 0$, and negative X at $t = 0.25$ T, and positive Y at $t = 0.5$ T, and positive X at $t = 0.75$ T, as shown in Fig. 4. The rotating induced current field can be perpendicular to cracks vertically at arbitrary scanning direction producing a larger distortion in magnetic field. The simulation results provide a strong evidence for the detection of cracks at arbitrary scanning direction using a double u-shaped orthogonal ACFM probe.

3   Cracks Detection Experiments

The double u-shaped orthogonal ACFM probe is set up, as shown in Fig. 5a. The excitation array is built by two orthogonal U-shaped MnZn ferrite yokes wound with 500 turns current-carry coils of 0.15 mm enameled copper wire each, according to theory model and FEM model. The experiment system is shown in Fig. 5b. The signal generator provides a sine signal with 6 kHz frequency and 1 V voltage as the initial driving signal. The initial driving signal and the orthogonal driving signal provided by the 90° phase shifter are used to drive the current-carrying coils of double u-shaped orthogonal ACFM probe. The detecting sensor picks up the disturbance magnetic signals caused by the disturbed uniform rotating alternating current filed. After the signal amplification, phase sensitive rectification and low pass filtering by

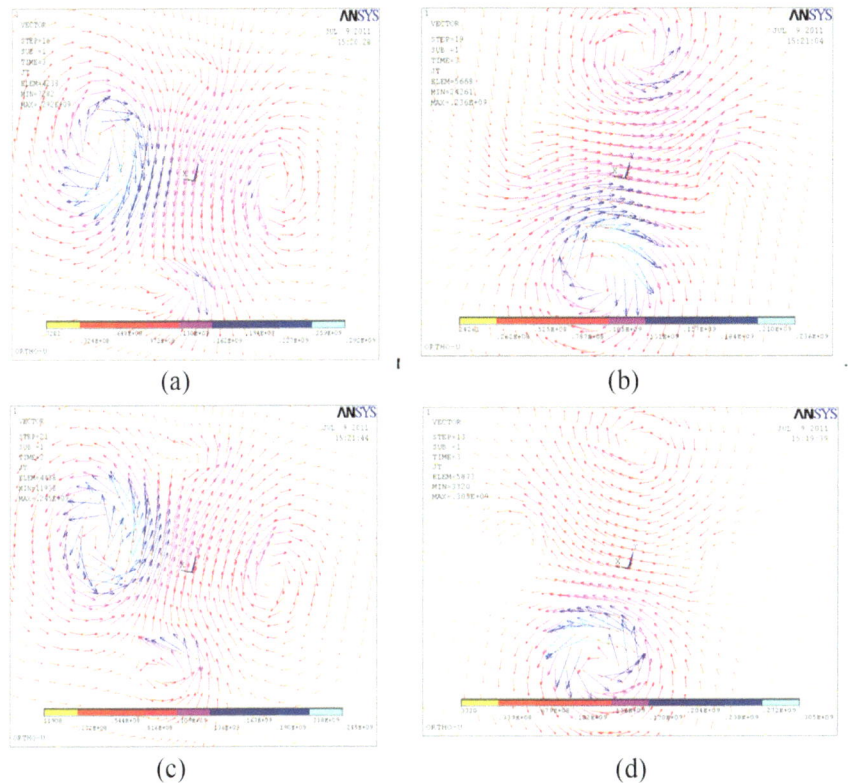

Fig. 4 The FEM simulation analysis results for induced AC field on the surface at different transient time, **a** t = 0, **b** t = 0.25T, **c** t = 0.5T, and **d** t = 0.75T

the condition module, these analog signals are transformed to digital signals and sent into the PC using the DAQ module. The intelligent identification software in PC will display the signals and identify arbitrary cracks with one pass scanning.

The specimen is a Q235 mild steel sheet with 45 mm length and 7 mm depth artificial rectangular crack. As shown in Fig. 5c, the double u-shaped orthogonal ACFM probe scans the crack from 0° to 90° with 10° increase simulating the detection of cracks at arbitrary scanning direction. The 0° angle indicates the double u-shaped orthogonal ACFM probe is parallel to the crack, while the 90° angle indicates the probe is perpendicular to the crack.

Figure 6a–d show the 0°, 30°, 60°, 90° angle cracks detection experimental results using the double u-shaped orthogonal ACFM probe. Comparing Fig. 5a with Fig. 1, it is clear that the B_X and B_Z signals are in accordance with the principle of ACFM, which proves the feasible of the experiment system. As shown in Fig. 6b, the perturbations of ACFM experiments results are still obvious at 30°. At 60°, as shown in

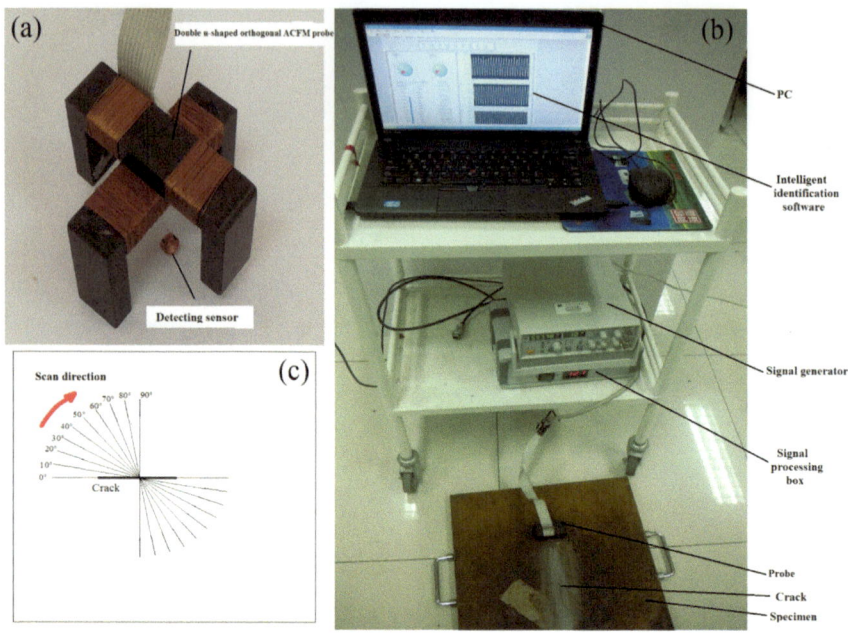

Fig. 5 The double u-shaped orthogonal ACFM probe. **a** The double u-shaped orthogonal ACFM probe and detecting sensor. **b** The experimental system. **c** Detection of cracks at arbitrary scanning direction

Fig. 6c, the B_X and B_Z are also according with the principle. Finally, when the scanning direction is perpendicular to the crack, as shown in Fig. 6d, the crack still can be recognized by the double u-shaped orthogonal ACFM probe perfectly.

The parameters for characterizing the inspection sensitivity, S_X and S_Z, are given as follows to reduce detection error and improve the SNR (Signal to Noise Ratio) [13].

$$\begin{cases} S_X = \frac{MX_{max}}{MX_0} \times 100\% \\ S_Z = \frac{MZ_{max}}{MX_0} \times 100\% \end{cases} \qquad (14)$$

where, MX_0 is the amplitude of B_X signal without crack, and MX_{max} and MZ_{max} are the maximum perturbations of B_X and B_Z caused by crack respectively.

As shown in Fig. 7, there is a little decrease of sensitivity by the double u-shaped orthogonal ACFM probe at different scanning detection. The maximum sensitivity of S_X is 32.6%, while the minimum sensitivity of S_X is 27.2%. The maximum decrease of sensitivity in S_X is 16.8%, which meets requirements of sensitive detection of cracks at arbitrary scanning direction. Meanwhile, the maximum sensitivity of S_Z is 69.1% and the minimum sensitivity of S_Z is 59.8%, whose maximum decrease is 13.5%. It suggests that the detection sensitivity of double u-shaped orthogonal ACFM probe changes a little for inspecting cracks at arbitrary scanning direction.

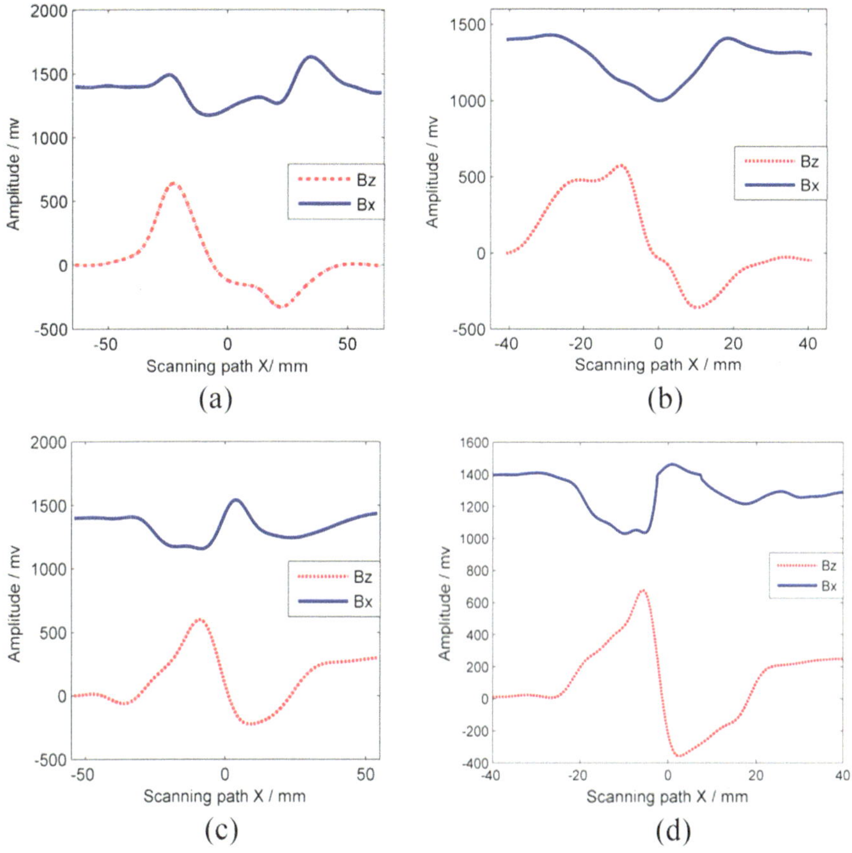

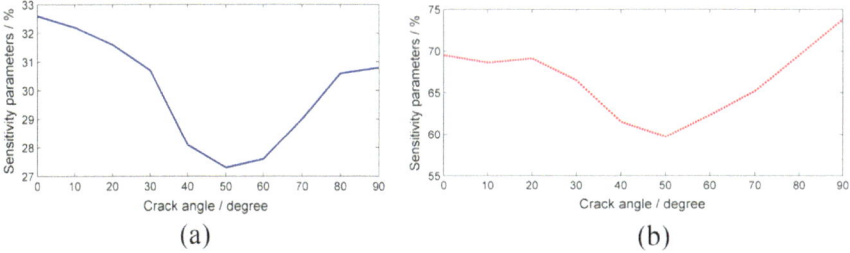

Fig. 6 B$_X$ and B$_Z$ signals from experiments using the double u-shaped orthogonal ACFM probe. **a** Crack detection at 0°. **b** Crack detection at 30°. **c** Crack detection at 60°. **d** Crack detection at 90°

Fig. 7 Detection sensitivity for cracks at different scanning direction. **a** Bx, **b** Bz

4 Conclusion

In this paper, a double u-shaped orthogonal ACFM probe is present for cracks detection at arbitrary scanning direction on structures. The induced rotating current field is analyzed by simulations. The cracks detection experiments at arbitrary scanning direction are carried out by the double u-shaped orthogonal ACFM probe test system. From these results, we conclude that the feasibility of double u-shaped orthogonal ACFM probe is verified by both EFM model and experiments. It is apparent that the double u-shaped orthogonal ACFM probe proposed in this paper can detect cracks effectively at arbitrary scanning direction.

References

1. W.D. Dover, R. Collins, D.H. Michael, The use of AC-field measurements for crack detection and sizing in air and underwater. Phil. Trans. R. Soc. Lond. A **320**, 271–283 (1986)
2. H. Rowshandel et al., A robotic approach for NDT of RCF cracks in rails using an ACFM sensor. Insight **53**(7), 368–376 (2011)
3. H. Saguy, D. Rittel, Application of ac tomography to crack identification. Appl. Phys. Lett. **91**, 084104 (2007)
4. H. Saguy, D. Rittel, Application of ac tomography to crack identification. Appl. Phys. Lett. **87**, 084103 (2005)
5. W. Li et al., A feed-through ACFM probe with sensor array for pipe string cracks inspection. NDT & E Int. **67**, 17–23 (2014)
6. M.P. Papaelias, et al., High-speed inspection of rolling contact fatigue in rails using ACFM sensors. Insight **51**(7), 366–369(4) (2009)
7. W. Li et al., Simulation analysis and experimental study of defect detection underwater by ACFM probe. China Ocean Eng. **27**, 277–282 (2013)
8. O. Yamashita, Effect of the surface integral in the torque equation of the electromagnetic angular momentum on the Faraday rotation. Opt. Commun. **284**, 4248 (2011)
9. Y.Z. He et al., Eddy current pulsed phase thermography and feature extraction. Appl. Phys. Lett. **103**, 084104 (2013)
10. Y. Lu, J.R. Bowler, T.P. Theodoulidis, An analytical model of a ferrite-cored inductor used as an eddy current probe. J. Appl. Phys. **111**, 103907 (2012)
11. C.K. Low, B.S. Wong, Defect evaluation using the alternating current field measurement technique. Insight **46**(10), 598–605(8) (2004)
12. W. Li et al., Analysis of the excitation frequency of a U-shaped ACFM system. NDT&E Int. **44**, 324–328 (2011)
13. G.L. Nicholson, C.L. Davis, Modelling of the response of an ACFM sensor to rail and rail wheel RCF cracks. NDT&E Int. **46**, 107–114 (2012)

A Novel Fatigue Crack Angle Quantitative Monitoring Method Based on Rotating Alternating Current Field Measurement

Abstract Under complicated fatigue loading conditions, cracks initiate nd grow in the arbitrary direction from corrosion pits in the aerospace equipment. The monitoring of crack propagation angle is very important for the safety assessment of the aerospace equipment, which is still a challenge by the conventional structural health monitoring (SHM) method. In this paper, a novel crack angle quantitative monitoring method is presented based on the rotating alternating current field measurement (RACFM). A theoretical model of the crack angle measuring method is established to analyze the perturbation principle of the induced electromagnetic field. The relationships between the angle, length and depth of the crack and the Bz signal are analyzed. The probe and testing system are established, and experiments are carried out. The results show that the phase of the Bz signal has a linear relationship with the crack angle for the same crack, and the amplitude of the Bz signal can correct the crack angle for the different cracks. The angle of fatigue cracks can be quantitatively measured by the Bz phase difference method based on the RACFM.

Keywords Fatigue crack · Angle quantitative monitoring · RACFM · Bz phase difference method · SHM

1 Introduction

Aluminum alloys are widely used in the aerospace industry due to their good specific properties such as high-thermal conductivity and high strength-to-weight ratio [1]. However, they are prone to localized corrosion pits due to corrosive environments such as plain water and saltwater [2, 3]. Because the aerospace structure is affected by cyclic loads, it is very easy to generate stress concentrations at the location of corrosion pits. Therefore, after corrosive pit formation, the next steps under complicated fatigue loading condition include pit growth, transition from pitting to fatigue crack initiation and short crack growth in arbitrary-angles [4–6]. A US Air Force airframe teardown analysis showed that 80% of fatigue crack initiation features are corrosion pits [7]. Short cracks expand along the depth and surface of the structure,

© The Author(s) 2024

X. Yuan et al., *Recent Development of Alternating Current Field Measurement Combine with New Technology*, https://doi.org/10.1007/978-981-97-4224-0_3

and then form typical surface cracks [8]. The entire structure may fail before the crack penetrates, which is concealed and hazardous. Therefore, it is necessary to inspect the process of pit-to-crack transition and the size and direction of fatigue crack growth for the safety assessment of the aerospace equipment.

The periodical non-destructive inspection of key parts is very important for the safety of equipment [9]. Nevertheless, conventional non-destructive testing methods to inspect structures regularly are generally expensive and labor-intensive. Moreover, it is not possible to repair and replace parts of the structure with cracks immediately, especially if the aircraft component is expensive. A new maintenance concept that is called structural health monitoring (SHM) technology has been proposed, which is to rationally allocate maintenance resources, and improve the safety and reliability of in-service aircraft structures [10]. Current SHM technologies mainly include GPS technology, optical fiber, strain gauges, accelerometers, acoustic emission, ultrasonic and electromagnetic. The GPS technology is used as an overall structure monitoring method, and it is difficult to monitor small cracks within large structures in time [11]. The optical fiber and strain gauges are commonly used techniques for monitoring structural stress and strain, but cannot monitor crack initiation and propagation [12, 13]. The accelerometers evaluate the structural state by measuring the mode of structural vibration, but the monitoring effect of nonlinear damage such as fatigue cracks needs to be improved [14]. The acoustic emission technology can detect dynamic cracks under external structural stress, but static cracks do not produce signals, and the application cost is high [15]. The ultrasound technology is not suitable for monitoring crack growth due to the need for couplants [16].

Electromagnetic monitoring methods are low-cost and easy-to-use based on electromagnetic induction and have good perspectives for use in the field of metallic structure health monitoring, which mainly include eddy current [17], alternating current potential drop [18] and metal magnetic memory (MMM) [19]. Li et al. measured the normal and tangential components of the stress-induced MMM signal by permanently installed magnetic sensor arrays [20]. MMM, that is susceptible to environmental interference and stress, is a weak magnetic signal which can easily cause misjudgment of crack monitoring results. Chaudhuri et al. proposed the alternating current potential drop method for weld toe fatigue crack initiation [21]. The alternating current potential drop requires alternating current to be applied to both ends of the structure, and complex array electrodes are placed on the surface of the structure for the crack monitoring. It destroys the anti-corrosion coating on the surface of the structures, and the monitoring results are greatly affected by human factors such as electrode installation. JENTEK Sensors Inc. has developed the Meandering Winding Magnetometer (MWM) sensor for monitoring of crack initiation and growth during fatigue tests and in service [22]. A rosette-like eddy current array sensor with high sensitivity was proposed for quantitatively monitoring hole-edge crack of aircraft structure [23, 24]. Sun et al. used one exciting coil covering the entire thickness and several sensing coils distributed along the axial length of the hole to quantitatively monitor a bolt hole crack in the radial and the axial directions [25]. The change-prone micro eddy current sensor was designed and fabricated with flexible

printed circuit board (FPCB) technology to monitor fatigue cracks of a metal structure [26]. The annular flexible eddy current array (FECA) sensor was developed for
quantitative monitoring of cracks in ferromagnetic steels under varying loads and
temperatures [27]. Yusa et al. designed the arrayed uniform eddy current probe for
crack monitoring and sizing of surface breaking cracks with the aid of a computational inversion technique [28]. However, the unidirectional eddy current generated
by the above method is only sensitive to cracks perpendicular to the direction of
the current. In addition, the traditional electromagnetic monitoring method cannot
obtain the direction of crack propagation.

In order to realize the detection of cracks at different angles, the RACFM technique
was proposed by different scholars. Hoshikawa et al. proposed a new ECT probe
called the Hoshi Probe which utilizes a uniform direction rotating eddy current to
reduce noises [29, 30]. Cracks of different angles are detected using the designed
probe, but it does not quantitatively analyze the relationship between the angle and
size of the crack and the amplitude and phase of the characteristic signal. Udpa et al.
designed a rotating current probe with sensor arrays to detect cracks at different
angles at a fastener site in layered structures [31, 32]. However, only the amplitude
of the characteristic signal is analyzed, the angle of the crack is obtained through
the C-scan image. All the above methods are non-destructive testing methods, but
neither of them involves the fixed-point monitoring of cracks.

In this paper, a novel fatigue crack monitoring method is proposed based on
the RACFM. The angle of crack is quantitative measured by the Bz (the magnetic
field perpendicular to the specimen) phase difference method. Firstly, two alternating
current signals with a phase difference of 90° and the same frequency are respectively
loaded on the orthogonal excitation coils, and the Bz signal is extracted by a sensor.
Secondly, the probe is placed in a position where the specimen has no defects, and the
real and imaginary components of the Bz signal are recorded as the calibration signal.
Then, the probe is placed at the corrosion pits, the real and imaginary component of
the crack is recorded as the monitoring signal. The differential signal is obtained by
subtracting the calibration signal from the monitoring signal. The real and imaginary
parts of the differential signal are converted into the amplitude and phase. Thirdly,
taking the 0° crack phase as the reference, the reference is subtracted from the Bz
phase of the measured crack to obtain the crack angle measurement result. Finally,
the crack angle measurement result is modified by the amplitude of the Bz signal. The
method proposed in this paper provides a new idea for monitoring the fatigue crack
direction of aluminum alloy materials in the aerospace industry. Compared with the
traditional eddy current monitering technology, the testing system is simpler and the
quantification accuracy of the crack angle measurement is improved.

The rest of the paper is organized as follows. The theoretical model of the RACFM
is established to analyze the influence of the angle of the crack on the distorted
electromagnetic field in Sect. 2. The influences of crack angles and sizes on the
characteristic signal are simulated by the 3D FEM model in Sect. 3. The RACFM
monitoring probe and testing system are developed and cracks of different angles,
lengths and depths are monitored in Sect. 4. Finally, the conclusions and future work
are outlined in Sect. 5.

2 Theoretical Model

The rotating uniform alternating current field can be induced using two orthogonal unidirectional coils with 90° phase shift alternating currents. The induced current J_{x1} and J_{x2} generated by the two excitation coils can be represent respectively as follows:

$$J_{x1}(t) = J_0 \sin(\omega t + \alpha_0) \vec{X}_1 \tag{1}$$

$$J_{y1}(t) = J_0 \cos(\omega t + \alpha_0) \vec{Y}_1 \tag{2}$$

where $\vec{X}_1$ and $\vec{Y}_1$ are unit vectors along the x and y axis respectively, J_0 is the amplitude of induced current density, α_0 is the phase of induced current, ω is the frequency of the induced current, and t is the time.

The directions of the two induced electric fields are orthogonal, their frequencies are equal and the phase difference is 90°, as shown in Fig. 1. When there are no defects, the induced electric field is uniform. When defects exist, the induced electric field is distorted.

As shown in Fig. 2, the induced currents are decomposed into the current component J_{x2} perpendicular to the crack direction and the current component J_{y2} parallel to the crack direction as follows:

$$J_{x2}(t) = (J_0 \sin(\omega t + \alpha_0) \cos(\theta) + J_0 \cos(\omega t + \alpha_0) \sin(\theta)) \vec{X}_2 \tag{3}$$

$$J_{y2}(t) = (J_0 \cos(\omega t + \alpha_0) \cos(\theta) - J_0 \sin(\omega t + \alpha_0) \sin(\theta)) \vec{Y}_2 \tag{4}$$

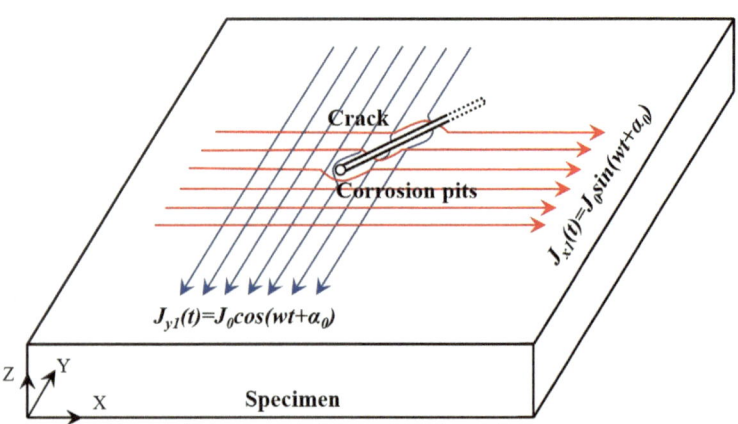

Fig. 1 Perturbations of uniform alternating current around corrosion pits

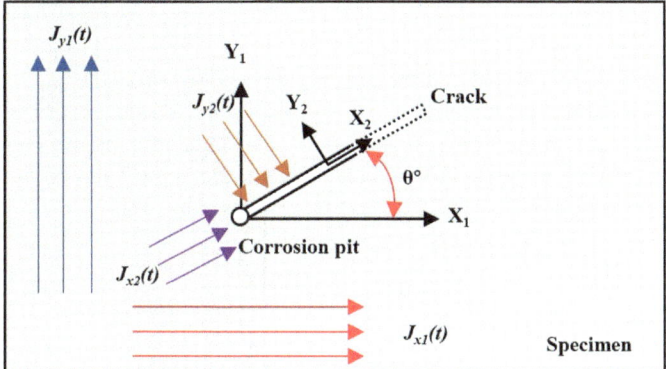

Fig. 2 Schematic diagram of induced current decomposition

where $\vec{X}_2$ and $\vec{Y}_2$ are unit vectors along the x_2 and y_2 axis respectively and θ is the crack angle.

According to the vector composition theorem, the total induced current density $J(t)$ in the specimen can be regarded as the superposition of two orthogonal induced currents J_{x2} and J_{y2}, and the total induced current amplitude $A_J(t)$ and phase angle $\theta_J(t)$ can be represent as follows [33]:

$$A_J(t) = \sqrt{J_{x2}(t)^2 + J_{y2}(t)^2} = J_0 \tag{5}$$

$$\theta_J(t) = \arctan\left(\frac{J_{x2}(t)}{J_{y2}(t)}\right) = \omega t + \alpha_0 + \theta \tag{6}$$

When there is no defect, the magnitude of the induced electric field generated by the rotating alternating current field is a fixed value. The direction of the induced electric field rotates periodically with time, and the rotation period is equal to the excitation current period. Furthermore, the phase of the induced electric field and the angle of the crack show a linear relationship.

The induced electric field is regarded as a number of straight wires carrying current, and the wire currents I_{x2} and I_{y2} in two directions can be expressed as follow:

$$I_{x2}(t) = (I_0 \sin(\omega t + \alpha_0) \cos(\theta) + I_0 \cos(\omega t + \alpha_0) \sin(\theta))\vec{X}_2 \tag{7}$$

$$I_{y2}(t) = (I_0 \cos(\omega t + \alpha_0) \cos(\theta) - I_0 \sin(\omega t + \alpha_0) \sin(\theta))\vec{Y}_2 \tag{8}$$

where I_0 is amplitude of the wire current.

At the no crack position, the magnetic induction intensity in the Z direction (Bz) is zero, since the magnetic field components of each long straight wire cancel each other out. When the crack exists, the current line at the tip of the crack forms a curved arc, so Bz is not zero.

The micro-current arc $I\mathrm{d}\tau$ on the current line near one tip of the crack is selected. The radius of the arc is r. According to the principle of electromagnetic field superposition, the integral of the magnetic field formed by the current micro-element along the current deflection path in the specimen must be along the Z direction. According to the Biot-Savart law, Bz_{x2} and Bz_{y2} can be expressed as [34]:

$$Bz_{x2}(t) = \oint \frac{\mu_0 I_{x2}(t)\mathrm{d}\tau}{4\pi\left(r^2+l^2\right)} \tag{9}$$

$$Bz_{y2}(t) = \oint \frac{\mu_0 I_{y2}(t)\mathrm{d}\tau}{4\pi\left(r^2+l^2\right)} \tag{10}$$

where μ_0 is vacuum permeability and l is lift-off.
Bz can be expressed as:

$$Bz(t) = Bz_{x2}(t) + Bz_{y2}(t) \tag{11}$$

As shown in Fig. 3, the induced current is much more significantly sensitive to vertical cracks than to parallel cracks, so $Bz_{x2}(t) \ll Bz_{y2}(t)$. Bz can be expressed as:

$$Bz(t) = \oint \frac{\mu_0 I_0 \cos(\omega t + \alpha_0 + \theta)\mathrm{d}\tau}{4\pi\left(r^2+l^2\right)} \tag{12}$$

It can be seen from (12) that at the crack tip, the angle of crack does not affect the amplitude of the Bz signal, but it has a linear relationship with the phase. Finally, the establishment process of the theoretical model is shown in Fig. 4.

Fig. 3 Schematic diagram of current disturbance. **a** Current lines are perpendicular to the crack; **b** Current lines are parallel to the crack

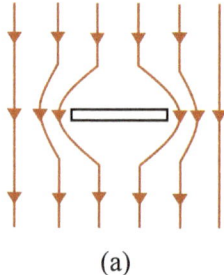

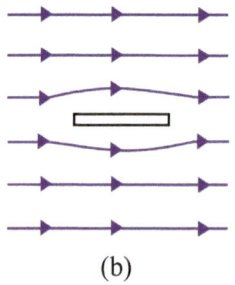

(a)   (b)

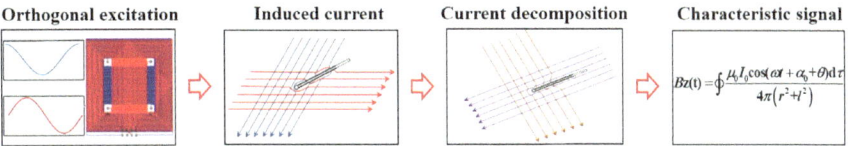

Fig. 4 Process of the theoretical model

3 Finite Element Analysis

3.1 Model Set Up

The 3D simulation model for the fatigue crack angle quantitative monitoring is built using the finite element software COMSOL, which is widely used to solve Maxwell's equations for modeling the electromagnetic field response due to its efficient computing performance and outstanding multi-field bidirectional direct coupling analysis capabilities [35, 36], as shown in Fig. 5. The model includes a specimen, coil-1, coil-2, crack, pick-up point and air. The lift-off of coil-1 is 0.8 mm and the lift-off of coil-2 is 0.5 mm. In order to ensure that the induced current is uniform in any direction, the two coils are loaded with different currents due to different lift-off heights. Coil-1 carries the alternating currents with a 0.31 A amplitude, 10 kHz frequency, and 0° phase. Coil-2 carries the alternating currents with a 0.3 A amplitude, 10 kHz frequency, and 90° phase. Two excitation coils are perpendicular to each other. The pick-up point is located at the coil center, whose lift-off distance is 1.5 mm. The thickness of the specimen is 10 mm, and the conductivity is 3.77 E7 S/m. The center of the excitation coils is at the tip of the crack. The mesh of the model adopts free tetrahedral. The grid size of specimen, coil-1, coil-2 is set to finer and air is set to fine in mesh module. A transient analysis is set for the model. The dimensions of the model and the characteristic parameters are shown in Table 1.

To explore the influence of corrosion pits on characteristic signals, cracks with and without corrosion pits are established as shown in Fig. 6. The length of the crack is 5 mm, the width is 0.2 mm, and the depth is 2 mm. The size of the corrosion pit is generally small, and the corrosion pit is spherical with a radius of 0.5 mm. The Bz signals of cracks with and without corrosion pits are shown in Fig. 7. The phase difference is 1.09°, and the amplitude difference is 3.18%. It shows that corrosion pits have little effect on the amplitude and phase of the Bz signal, which is because the boundary of the corrosion pit is smoother than the tip of the crack, and the current is more likely to gather at the tip of the crack. So, the following simulations and experiments adopt a model without corrosion pits to simplify the study.

As shown in Fig. 8, the induced currents at different moments are extracted. It can be seen from the figure that the angle of the induced current change between two adjacent times is about 45°, and the induced current rotates counterclockwise. So, the structure of the orthogonal excitation coils can induce a periodic rotating uniform

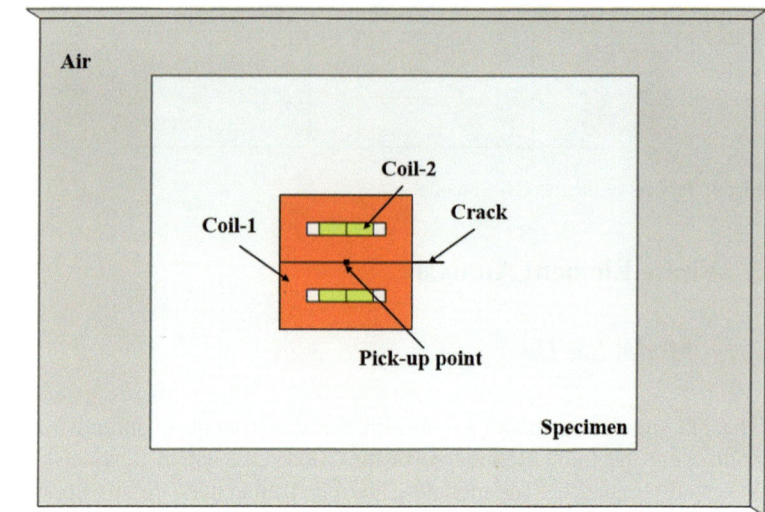

Fig. 5 Fatigue crack angle quantitative monitoring FEM model

Table 1 Parameters of the model

Name	Value
Specimen	Aluminum, length = 200 mm, width = 150 mm, depth = 10 mm
Air	Length = 300 mm, width = 200 mm, depth = 75 mm
Coil-1	Length = 54 mm, width = 54 mm, coil turns = 32, diameter = 2 mm, Symmetry axis = X
Coil-2	Length = 54 mm, width = 54 mm, coil turns = 32, diameter = 2 mm, Symmetry axis = Y

Fig. 6 FEM models of cracks with and without corrosion pits

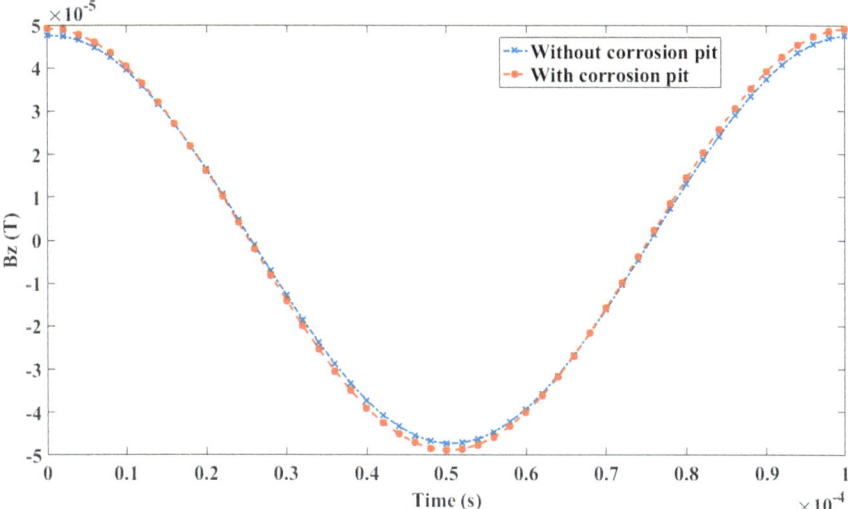

Fig. 7 Simulation results of cracks with and without corrosion pits

alternating current on the surface of the specimen, and the induced current will be disturbed when it encounters a crack.

3.2 *Characteristic Signal Analysis of Cracks with Different Angles*

To explore the correlations between the crack angle and the Bz signal, the crack (length = 20 mm, width = 0.2 mm, depth = 2 mm) is established, and the cracks with different angles (0°, 60°, 120°, 180°, 240°, 300°) are simulated. The Bz signals at the tip of crack are extracted, as shown in Fig. 9a. It can be seen from the figure that the Bz signals of cracks at different angles have the same amplitude and different phases. Hence, the Bz phases of the cracks at different angles are extracted, as shown in Fig. 9b. It shows that the phase of the Bz signal has a linear relationship with the angle of the crack. This is because the orthogonal excitation coils generate a rotating uniform alternating current in the specimen and the direction of the initial distorted current has a linear relationship with the angle of the crack.

Taking the Bz phase of the 0° crack as the reference, the reference is subtracted from the Bz phases of the cracks at all angles, and crack angle measurement results are obtained, as shown in Fig. 9c. It shows that, in the polar diagram, the distance from the signal point to the origin is the amplitude of the Bz signal, and the phase of the signal point is the measured angle of the crack.

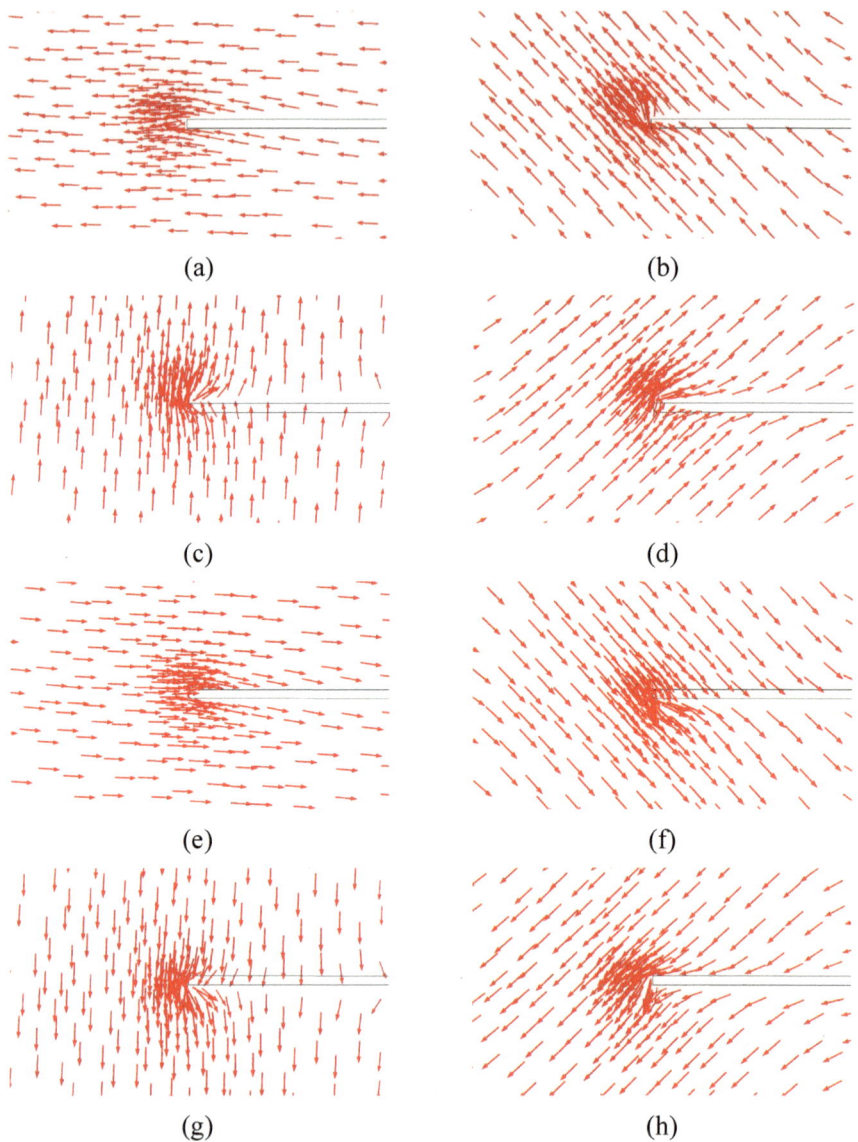

Fig. 8 Results of induced current at different transient times. **a** T/8; **b** 2 T/8; **c** 3 T/8; **d** 4 T/8; **e** 5 T/8; **f** 6 T/8; **g** 7 T/8; **h** 8 T/8

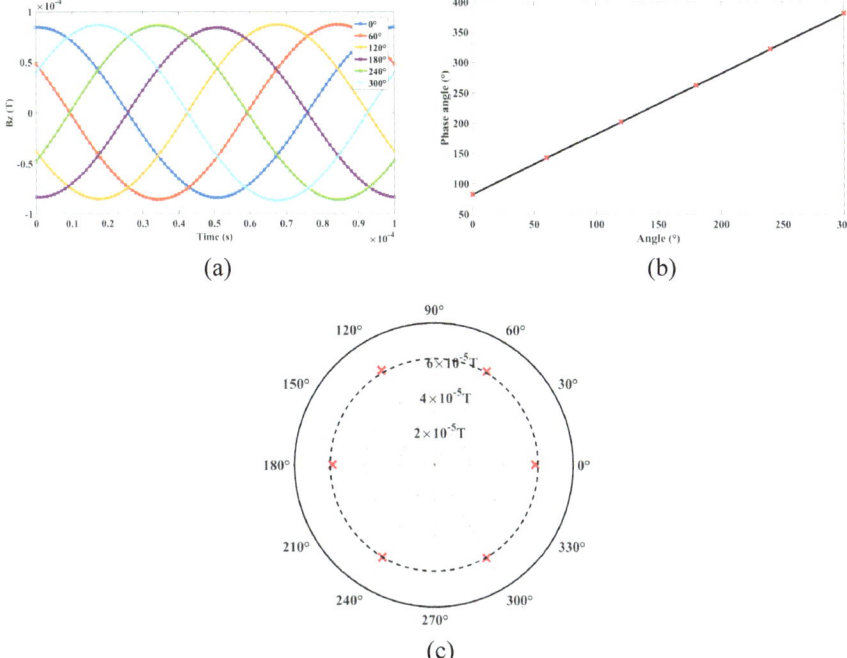

Fig. 9 Characteristic signals of cracks with different angles. **a** Bz signals of cracks with different angles; **b** Relationships between the crack angle and the phase of the Bz signal; **c** Crack angle measurement results

3.3 Characteristic Signal Analysis of Cracks with Different Lengths and Depths

In order to explore the correlations between the size of the crack and the Bz signal, multiple angles (0°, 60°, 120°, 180°, 240°, 300°) of cracks with different lengths (2 mm, 4 mm, 6 mm, 8 mm, 10 mm) and different depths (1 mm, 2 mm, 3 mm, 4 mm, 5 mm) are simulated to determine the angle quantification algorithm for cracks of different sizes. Taking the 0° crack phase (length = 20 mm, width = 0.2 mm, depth = 2 mm) as the reference, according to the method in Sect. 3.3, the polar diagrams of different length and depth cracks are obtained, as shown in Fig. 10a and Fig. 10b. The amplitude of the Bz signal is mainly affected by the size of the crack. The longer the crack length or the deeper the crack depth, the greater the amplitude of the Bz signal. This is because the increase in the size of the crack leads to an increase in the perturbation current density. The phase of the Bz signal is mainly affected by the angle of the crack. For the same crack, the phase difference of the Bz signal is equal to the angle difference of that crack taken in any two directions.

However, the angle measurement results of cracks with the same angle and different sizes have certain deviations by the above method. It can be seen from

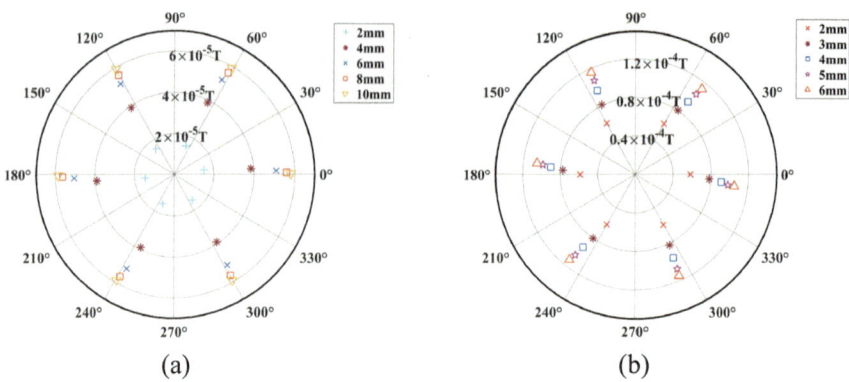

(a)   (b)

Fig. 10 Characteristic signal of cracks with different lengths and depths. **a** Polar diagram of cracks with different lengths; **b** Polar diagram of cracks with different depths

Fig. 10a, b that the amplitude of the Bz signal and the crack angle measurement result show a certain correlation. When the amplitude of the Bz signal increases, the result of the angle measurement will be smaller, so the amplitude of the Bz signal can be used to modify the measured angle of the crack.

4 Experimental Setup and Result

4.1 Probe and System Setup

The RACFM monitoring probe is built, as shown in Fig. 11. It includes two planar excitation coils and a magnetic sensor. The length and width of the excitation coils are 57 mm and 59 mm, respectively. The two coils are designed to be perpendicular to induce a periodic rotating uniform alternating current. Each planar excitation coil is composed of two symmetrically distributed coils, and the winding directions of the coils are clockwise and counterclockwise respectively. The number of turns of each coil is 32, the width of the wire is 0.2 mm, and the pitch of the wire is 0.15 mm. The magnetic field sensor is located in the center above the excitation coils, which is the commercial TMR packages type 2301 from Duowei in China.

The RACFM testing system is developed, as shown in Fig. 12, which includes a signal generator, a power amplifier, a specimen, a low pass filter module, a signal amplifier module, a lock-in amplifier module, a computer and a capture card. Two sinusoidal signals with 90° phase shift are generated by a signal generator and amplified by a power amplifier. Two amplified signals are loaded on two orthogonal excitation coils respectively. The periodic rotating electromagnetic field is generated by the excitation coils. The TMR sensor picks up the induced magnetic field and converts it into a voltage signal. The voltage signal is filtered and amplified. The DC output

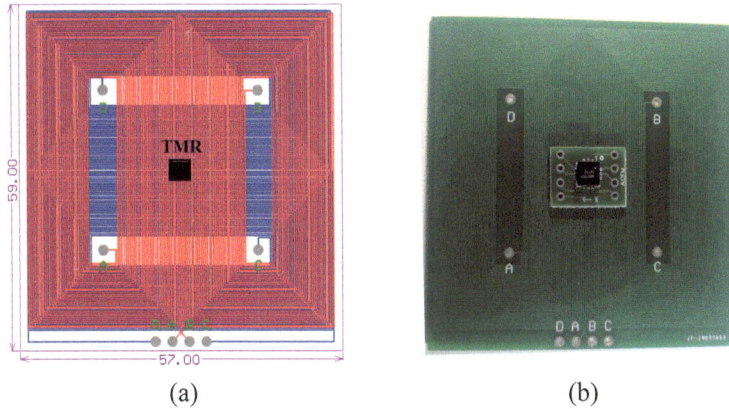

Fig. 11 Design of probe. **a** Schematic diagram of probe; **b** Picture of probe

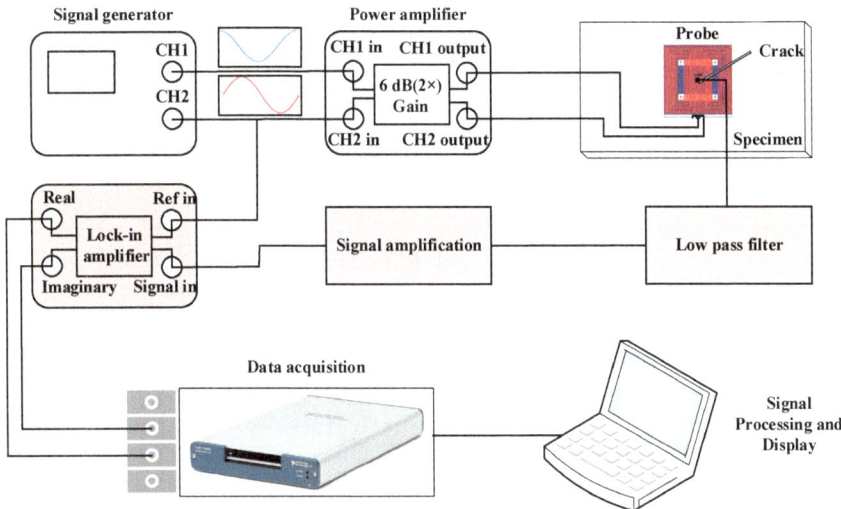

Fig. 12 Block diagram of experimental system

of the lock-in amplifier is acquired by the DAQ card. These signals can be processed and analyzed by an intelligent software developed in LabVIEW and MATLAB [37].

4.2 Crack Length Monitoring

The first specimen tested in this paper is an aluminum plate with five 3 mm deep and 0.2 mm wide cracks. Artificial cracks of different length are machined into the

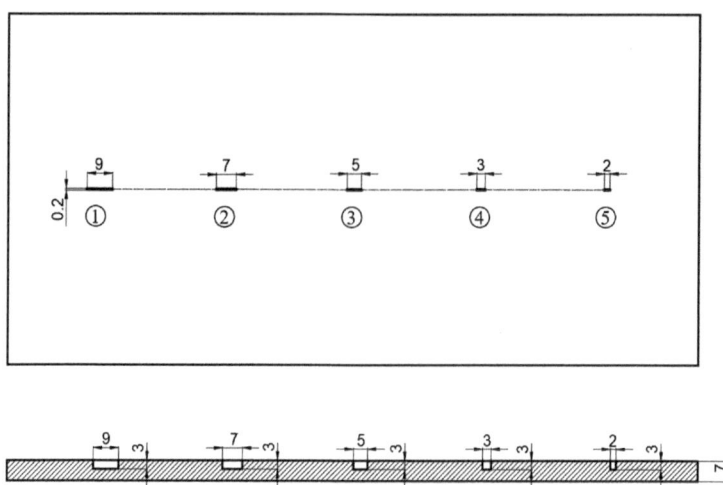

Fig. 13 Specimen with different length cracks

plate by electrical discharge machining (EDM) method. The lengths of the cracks are 9 mm, 7 mm, 5 mm, 3 mm and 2 mm, as shown in Fig. 13.

The established testing system is used for monitoring the cracks of different lengths. The excitation frequency is 10 kHz, and the voltage is 3 V. Firstly, the probe is placed in a position where the specimen has no defects, and the real and imaginary components of the Bz signal are recorded as the calibration signal. Secondly. the probe is placed at the tip of the crack, and the probe is rotated counterclockwise from 0°. At the same time, the real and imaginary components of the cracks of different length are recorded as the monitoring signal. The differential signal is obtained by subtracting the calibration signal from the monitoring signal. Thirdly, the real and imaginary parts of the differential signal are converted into the amplitude and phase. Finally, taking the 0° crack phase (length = 9 mm, width = 0.2 mm, depth = 3 mm) as the reference, according to the method in Sect. 3.3, the angle measurement results of different length cracks is obtained, as shown in Fig. 14. It can be seen from the figure that when the probe rotates about the centre of the circle relative to the crack tip once, the phase and amplitude of the Bz signal also behave in the same manner, and the phase angle changes by 360°, which proves that the phase of the Bz signal can be used to measure the angle of the crack. What's more, when the crack length is long, the graph drawn by the phase and amplitude is close to a circle, and when the crack is short, the graph drawn by the phase and amplitude is elliptical. This is because the induced current in the specimen is not ideally uniform in any direction. When the crack length is short, the distance between the two tips of the crack is closer, the perturbation electromagnetic field generated at the other tip will affect the amplitude of the response signal resulting in large changes in the Bz amplitude of cracks at different angles.

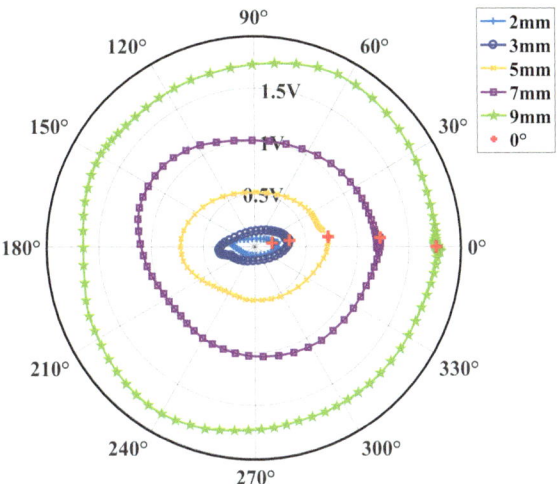

Fig. 14 Testing results of different length cracks

In order to analyze the relationship between the amplitude of the Bz signal and the length of the crack, the amplitudes of the Bz signals of the different length cracks with an angle of $0°$ are extracted, as shown in Fig. 15. It can be seen from the figure that as the length of the crack expands, the amplitude of the Bz signal also increases, and the quadratic function is used fit the relationship between the crack length and the Bz amplitude, as shown in Eq. (13).

$$A = 0.01L^2 + 0.12L - 0.10 \tag{13}$$

where L is the length of crack and A is amplitude of Bz signal.

4.3 Crack Depth Monitoring

The second specimen tested in this paper is an aluminum plate with five 16 mm long and 0.5 mm wide cracks. The thickness of the plate is 9 mm. The depths of the cracks are 9 mm, 7 mm, 5 mm,3 mm and 1 mm, as shown in Fig. 16.

According to the above experimental steps, the cracks of different depth are measured, and the amplitudes and phases are obtained as shown in Fig. 17. The phase angle of the Bz changes with the rotation of the probe, and the amplitude and phase of the Bz form a complete circle. Compared with cracks of different lengths, the depth of the cracks has no obvious effect on the Bz amplitude of different angles. But relative to No.5, polar diagrams with different depth cracks are more elliptical, because the width of the crack becomes larger, and the tip effect of the rectangular crack has an impact on the amplitude of each angle.

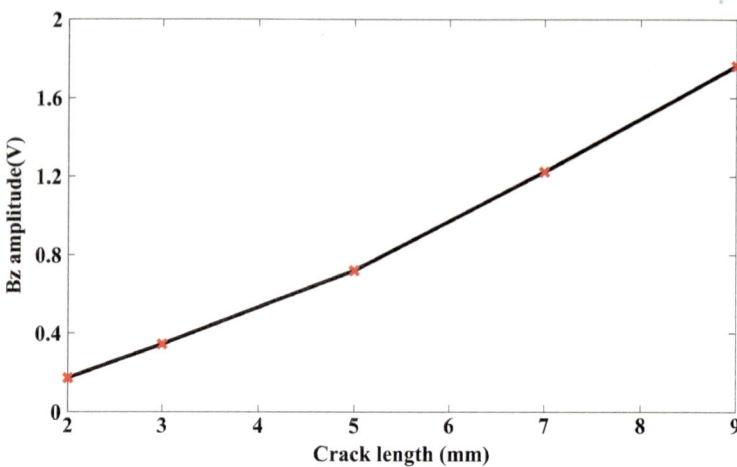

Fig. 15 Relationship between crack length and Bz amplitude

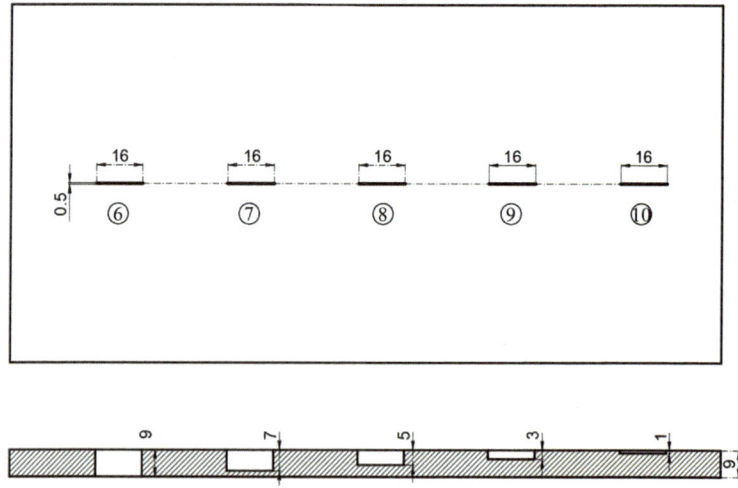

Fig. 16 Specimen with different depth cracks

In order to analyze the relationship between the amplitude of the Bz signal and the depth of the crack, the amplitudes of the Bz signals of the different depth cracks with an angle of $0°$ are extracted, as shown in Fig. 18. It can be seen from the figure that as the depth of the crack expands, the amplitude of the Bz signal also increases, and the quadratic function is used fit the relationship between crack depth and Bz amplitude, as shown in Eq. (14).

$$A= -0.04D^2 + 1.16D+0.05 \tag{14}$$

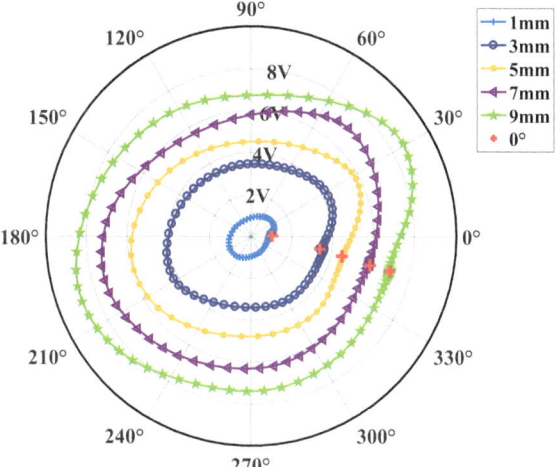

Fig. 17 Testing results of different depth cracks

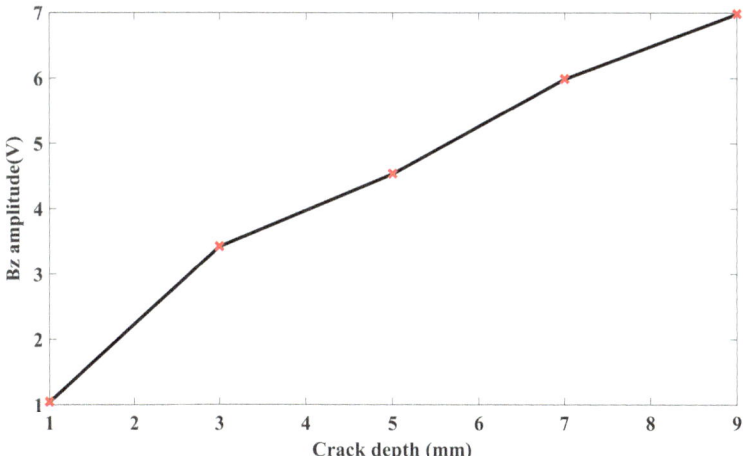

Fig. 18 Relationship between crack depth and Bz amplitude

where D is the depth of crack and A is amplitude of the Bz signal.

4.4 Modification of the Measured Angle of the Crack

According to Sect. 3.3, it can be seen that the crack angle measurement result has a certain deviation, which is also verified by the experimental results. As shown in Figs. 14 and 17, the curves of cracks of different sizes do not all start from 0° in the polar diagram. So further modification is required to obtain a more accurate angle measurement result. The angle measurement result of different size cracks with an angle of 0° are shown in Table 2. It can be seen from the table that the maximum error of the crack angle is 13.77°. This is because the angle measurement result is also affected by the crack size, which leads to inaccuracy in the measured angle of the crack.

The amplitudes of the Bz signal and the angle measurement results are extracted as shown in Fig. 19. It shows that when the amplitude of the Bz signal becomes larger, the angle measurement result becomes smaller, which is consistent with the simulation results. Therefore, the angle measurement result needs to be modified by the amplitude of the Bz signal. The quadratic function is used fit the relationship between the amplitude and the angle measurement result, as shown in Eq. (15).

$$P= 0.73A^2 - 8.85A+12.68 \qquad (15)$$

where P is the angle measurement result. and A is the amplitude of the Bz signal.

According to Eq. (15), the angle measurement results are modified using the amplitudes of the Bz signal, and the modified measured angles of the crack are shown in Table 3. And the error of crack angle measurement before and after the modification is compared, as shown in Fig. 20. It shows that the measurement error of the crack angle is significantly reduced after the modification, and the maximum error

Table 2 Angle measurement results of different size cracks

Crack	1	2	3	4	5	6	7	8	9	10
Error	–	3.94°	7.45°	10.07°	11.16°	−13.81°	−13.30°	−11.94°	−9.86°	1.79°

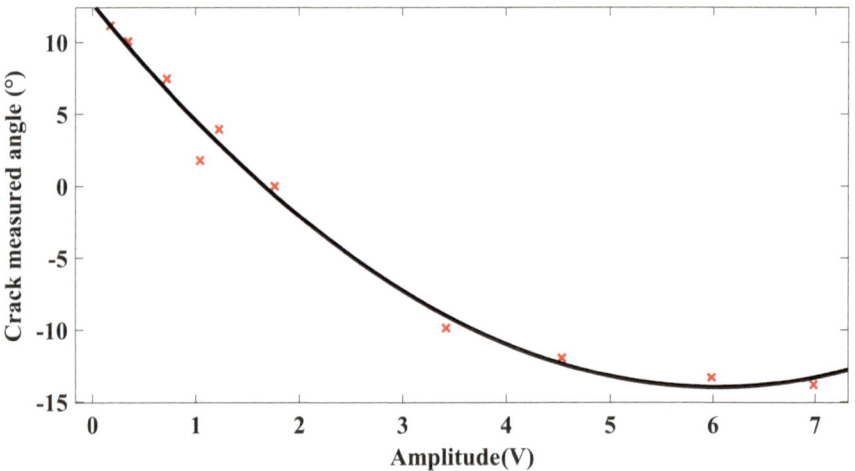

Fig. 19 Relationship between phase and amplitude

Table 3 Modified measured angles of the crack

Crack	1	2	3	4	5	6	7	8	9	10
Error	–	−0.34°	−0.11°	0.31°	0.67°	0.94°	−0.19°	0.15°	1.46°	3.11°

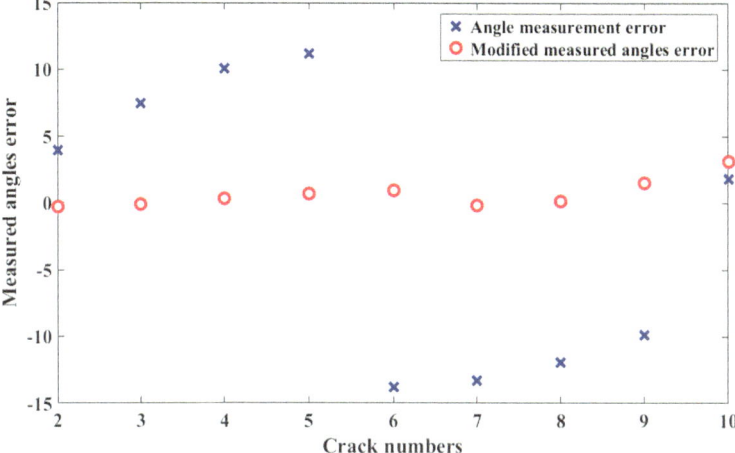

Fig. 20 Comparison of crack angle measurement

is reduced from 13.81° to 3.11°. It shows that this method can realize quantitative monitoring of the angle by the corrosion pit.

5 Conclusions and Further Work

In this work, a novel fatigue crack angle quantitative monitoring method is presented based on the RACFM, which measures the angle of the crack through the corrosion pit. The main research results are shown as follows:

- The theoretical model of the RACFM is established to analyze the response principle of the induced uniform electromagnetic field on the plate and the disturbance of the induced current by different crack angles. It indicates that, at the crack tip, the angle of crack does not affect the amplitude of the Bz signal, and it has a linear relationship with the phase.
- The 3D simulation model of the fatigue crack angle quantitative monitoring is built. When the sinusoidal signals with initial phases of 0° and 90° are loaded on the orthogonal excitation coils, a periodic rotating uniform current field is induced in the surface of the specimen. The center of the excitation coil is placed at the tip of the crack. The relationship between the angle, length, depth of the crack and the amplitude, phase of the Bz signal is analyzed. For the Bz signals of cracks

with the same angle and different sizes, the larger the Bz amplitude, the smaller the crack angle measurement result.

- The RACFM monitoring probe, which includes two planar excitation coils and a TMR sensor, and testing system are developed. Cracks with different angles, lengths and depths are measured. The amplitude and phase of the Bz differential signal are obtained and used to draw polar diagrams. The results show that the phase of the Bz signal can be used to quantify the angle of the crack. The amplitude of the Bz signal and the length and depth of the crack are in a quadratic function respectively, and the amplitude of the Bz signal can be used to modify the crack angle measurement result.

The experimental results suggest that the proposed method has the potential to quantitatively monitor crack propagation angle. In addition, the traditional eddy current monitoring technology only extracts the amplitude of the characteristic signal, and the accuracy of the crack angle monitoring depends on the density of the sensor. The proposed the *Bz* phase difference method utilizes the amplitude and phase of the detection signal, and the crack angle can be measured at one end of the crack, which greatly improves the quantification accuracy of the crack angle measurement. And the use of orthogonal excitation can eliminate the influence of crack angle on sensitivity. It should be noted that there are some limitations of the proposed method. Most aerospace equipment is a curved structure, and the influence of curvature on the induction signal is not considered in this paper. However, the development of flexible PCB technology provides an effective method to solve this problem. The stability and reliability of the detection system need to be improved if it is to be implemented in real case. Further work will focus on the monitoring of bolt hole cracks in multilayer structures.

References

1. N.E.C. Co, J.T. Burns, Effects of macro-scale corrosion damage feature on fatigue crack initiation and fatigue behavior. Int. J. Fatigue **103**, 234–247 (2017). https://doi.org/10.1016/j.ijfatigue.2017.05.028
2. V. Sabelkin, H.E. Misak, V.Y. Perel, S. Mall, Crack initiation from corrosion pit in three aluminum alloys under ambient and saltwater environments. J. Mater. Eng. Perform. **25**(4), 1631–1642 (2016). https://doi.org/10.1007/s11665-016-1996-5
3. V. Sabelkin, V.Y. Perel, H.E. Misak, E.M. Hunt, S. Mall, Investigation into crack initiation from corrosion pit in 7075–T6 under ambient laboratory and saltwater environments. Eng. Fract. Mech. **134**, 111–123 (2015). https://doi.org/10.1016/j.engfracmech.2014.12.016
4. K. van der Walde, B.M. Hillberry, Initiation and shape development of corrosion-nucleated fatigue cracking. Int. J. Fatigue **29**(7), 1269–1281 (2007). https://doi.org/10.1016/j.ijfatigue.2006.10.010
5. V. Sabelkin, S. Mall, H. Misak, Investigation into corrosion pit-to-fatigue crack transition in 7075–T6 aluminum alloy. J. Mater. Eng. Perform. **26**(6), 2535–2541 (2017). https://doi.org/10.1007/s11665-017-2697-4

6. P.S. Pao, S.J. Gill, C.R. Feng, On fatigue crack initiation from corrosion pits in 7075-T7351 aluminum alloy. Scripta Mater. **43**(5), 391–396, 2000–01–01 (2000). https://doi.org/10.1016/S1359-6462(00)00434-6

7. G.A. Shoales, S.A. Fawaz, M.R. Walters, Compilation of damage findings from multiple recent teardown analysis programs, in *Bridging the Gap between Theory and Operational Practice, 25th ICAF Symposium* (Springer, Rotterdam, The Netherlands, 2009), pp. 187–207.

8. Q. Wang, Effect of pitting corrosion on very high cycle fatigue behavior. Scripta Mater. **49**(7), 711–716 (2003). https://doi.org/10.1016/S1359-6462(03)00365-8

9. X. Yin, J. Fu, W. Li, G. Chen, D.A. Hutchins, A capacitive-inductive dual modality imaging system for non-destructive evaluation applications. Mech. Syst. Signal Pr. **135**, 106403 (2020). https://doi.org/10.1016/j.ymssp.2019.106403

10. H.A. Sodano, Development of an automated eddy current structural health monitoring technique with an extended sensing region for corrosion detection. Struct. Health Monit. **6**(2), 111–119 (2007). https://doi.org/10.1177/1475921706072065

11. G.W. Roberts, X. Tang, C.J. Brown, Measurement and correlation of displacements on the Severn suspension bridge using GPS. Appl. Geomatics **11**(2), 161–176 (2019). https://doi.org/10.1007/s12518-018-00251-6

12. R. Min, Z. Liu, L. Pereira, C. Yang, Q. Sui, C. Marques, Optical fiber sensing for marine environment and marine structural health monitoring: a review. Opt. Laser Technol. **140**, 107082 (2021). https://doi.org/10.1016/j.optlastec.2021.107082

13. D. Zhao et al., Development of an embedded thin-film strain-gauge-based SHM network into 3D-woven composite structure for wind turbine blades, in *SPIE*, vol. 10171, p. 101710C (2017). https://doi.org/10.1117/12.2259808

14. S. Kavitha, R. Joseph Daniel, K. Sumangala, design and analysis of MEMS comb drive capacitive accelerometer for SHM and seismic applications. Measurement **93**, 327–339 (2016). https://doi.org/10.1016/j.measurement.2016.07.029

15. P. Feng, P. Borghesani, H. Chang, W.A. Smith, R.B. Randall, Z. Peng, Monitoring gear surface degradation using cyclostationarity of acoustic emission. Mech. Syst. Signal Pr. **131**, 199–221 (2019). https://doi.org/10.1016/j.ymssp.2019.05.055

16. M. Matsushita, N. Mori, S. Biwa, Transmission of Lamb waves across a partially closed crack: numerical analysis and experiment. Ultrasonics **92**, 57–67 (2019). https://doi.org/10.1016/j.ultras.2018.09.007

17. V. Zilberstein, MWM eddy-current arrays for crack initiation and growth monitoring. Int. J. Fatigue **25**(9–11), 1147–1155 (2003). https://doi.org/10.1016/j.ijfatigue.2003.08.010

18. M.C. Lugg, The analysis of sparse data in ACPD crack growth monitoring. Ndt&E Int. **21**(3), 153–158 (1988). https://doi.org/10.1016/0308-9126(88)90446-4

19. Z.C. Qiu, H. Li, Z.J. Yao, J.S. Yang, R.L. Zhang, Early prediction of fracturing failure for bolted joints based on enhanced metal magnetic memory. Insight **61**(10), 603–607 (2019). https://doi.org/10.1784/insi.2019.61.10.603

20. Z. Li, S. Dixon, P. Cawley, R. Jarvis, P.B. Nagy, S. Cabeza, Experimental studies of the magneto-mechanical memory (MMM) technique using permanently installed magnetic sensor arrays. Ndt&E Int. **92**, 136–148 (2017). https://doi.org/10.1016/j.ndteint.2017.07.019

21. S. Chaudhuri, J. Crump, P.A.S. Reed, B.G. Mellor, High-resolution 3D weld toe stress analysis and ACPD method for weld toe fatigue crack initiation. Weld World **63**(6), 1787–1800 (2019). https://doi.org/10.1007/s40194-019-00792-3

22. V. Zilberstein, D. Schlicker, K. Walrath, V. Weiss, N. Goldfine, MWM eddy current sensors for monitoring of crack initiation and growth during fatigue tests and in service. Int. J. Fatigue **23**, 477–485 (2001). https://doi.org/10.1016/S0142-1123(01)00154-2

23. G. Chen, W. Zhang, Z. Zhang, X. Jin, W. Pang, A new rosette-like eddy current array sensor with high sensitivity for fatigue defect around bolt hole in SHM. Ndt&E Int **94**, 70–78 (2018). https://doi.org/10.1016/j.ndteint.2017.12.001

24. P. Li, L. Cheng, Y. He, S. Jiao, J. Du, H. Ding, J. Gao, Sensitivity boost of rosette eddy current array sensor for quantitative monitoring crack. Sens. Actuators, A **246**, 129–139 (2016). https://doi.org/10.1016/j.sna.2016.05.023

25. H. Sun, T. Wang, Q. Liu, X. Qing, A novel eddy current array sensing film for quantitatively monitoring hole-edge crack growth in bolted joints. Smart Mater. Struct. **28**(1), 15018 (2018). https://doi.org/10.1016/j.ijfatigue.2017.05.028

26. S. Jiao, L. Cheng, X. Li, P. Li, H. Ding, Monitoring fatigue cracks of a metal structure using an eddy current sensor. EURASIP J. Wirel. Commun. Netw. **2016**(1) (2016). https://doi.org/10.1186/s13638-016-0689-y

27. Y. Song, T. Chen, R. Cui, Y. He, K. Ding, Annular flexible eddy current array (A-FECA) sensor for quantitative monitoring of cracks in ferromagnetic steels under varying loads and temperatures. Meas. Sci. Technol. **31**(12) (2020). https://doi.org/10.1088/1361-6501/aba8b0

28. N. Yusa, H. Hashizume, R. Urayama, T. Uchimoto, T. Takagi, K. Sato, An arrayed uniform eddy current probe design for crack monitoring and sizing of surface breaking cracks with the aid of a computational inversion technique. Ndt&E Int **61**, 29–34 (2014). https://doi.org/10.1016/j.ndteint.2013.09.004

29. H. Hoshikawa, K. Koyama, A new ECT probe with rotating direction eddy current, in *Review of Progress in QNDE*, vol. 15A (Plenum Press, New York, 1995), , pp 1091–1098

30. H. Hoshikawa, K. Koyama, A new eddy current probe using uniform rotating eddy currents. Mater. Eval. **56**(1), 85–89 (1998)

31. C. Ye, Y. Huang, L. Udpa, S.S. Udpa, Differential sensor measurement with rotating current excitation for evaluating multilayer structures. IEEE Sens. J. **16**(3), 782–789 (2016). https://doi.org/10.1109/JSEN.2015.2488289

32. G. Yang, G. Dib, L. Udpa, A. Tamburrino, S.S. Udpa, Rotating field EC-GMR sensor for crack detection at fastener site in layered structures. IEEE Sens. J. **15**(1), 463–470 (2015). https://doi.org/10.1109/JSEN.2014.2341653

33. W. Li, X. Yuan, G. Chen, J. Ge, X. Yin, K. Li, High sensitivity rotating alternating current field measurement for arbitrary-angle underwater cracks. NDT E Int. NDT&E Int. **79**, 123–131 (2016). https://doi.org/10.1016/j.ndteint.2016.01.003

34. X. Yuan, W. Li, G. Chen, X. Yin, X. Li, J. Liu, J. Zhao, J. Zhao, Visual and intelligent identification methods for defects in underwater structure using alternating current field measurement technique. IEEE Trans. Industr. Inf. **18**(6), 3853–3862 (2022). https://doi.org/10.1109/JSEN.2014.2341653

35. Multiphysics, COMSOL. v. 5.4; COMSOL AB: Stockholm, Sweden (2018)

36. T. Chen, Y. He, J. Du, A High-sensitivity flexible eddy current array sensor for crack monitoring of welded structures under varying environment. Sensors-Basel **18**(6), 1780 (2018). https://doi.org/10.3390/s18061780

37. MATLAB. v. R2021b; MathWorks: Massachusetts, USA (2021)

Inspection of Both Inner and Outer Cracks in Aluminum Tubes Using Double Frequency Circumferential Current Field Testing Method

Abstract Aluminum and alloy tubes are widely used in industrial fields because of the advantage of good corrosion resistance, high thermal conductivity and light weight. Due to the stress corrosion cracking (SCC) and fatigue corrosion cracking (FCC), both inner and outer cracks generates in the aluminum tube. It is still a challenge to inspect all inner and outer surface cracks in the thick-wall aluminum tube in real time by one scan using the nondestructive testing (NDT) method. A double frequency circumferential current field testing (CCFT) method is presented for the inspection of both inner and outer cracks in the aluminum tube in a one pass scan. A simulation model is proposed to extract characteristic signals of inner and outer cracks at two excitation frequencies. The bobbin-type probe is developed and excited by the synthetic double frequency excitation signal. Both inner and outer cracks are tested by the double frequency CCFT system. Results show that both inner and outer cracks can be identified, distinguished and evaluated by the double frequency CCFT method in a one pass scan.

Keywords Inner and outer cracks · Double frequency · Circumferential current field testing · Aluminum tube · Inspection

1 Introduction

Aluminum and alloy tubes are widely used in power generation, transportation, construction, automobile and aerospace industries due to the advantage of good corrosion resistance, high thermal conductivity, light weight and high strength [1–3]. However, most of aluminum tubes suffer from severe environments, such as corrosive medium, time-varying thermal stress and heavy load stress. The longitudinal stress corrosion cracking (SCC) and fatigue corrosion cracking (FCC) are easily appeared and extended in both inner and outer surfaces of the aluminum tube. The SCC and FCC grow and gather rapidly as the complex stress and corrosion, which brings potential security problems [4, 5].

© The Author(s) 2024

X. Yuan et al., *Recent Development of Alternating Current Field Measurement Combine with New Technology*, https://doi.org/10.1007/978-981-97-4224-0_4

Thus the nondestructive testing (NDT) techniques should be carried out to inspect the crack and evaluate the condition of the aluminum tube regularly [6, 7]. However, in most case the costs of time and NDT operations are expensive using two techniques or multiple scans to test both inner and outer surface of the aluminum tube respectively during the in-service time. Thus it is of prime significance to propose an effective NDT method to detect both inner and outer cracks in a one pass scan.

The magnetic flux leakage (MFL) testing method is usually employed to detect inner and outer defects in the ferromagnetic tube, such as the well-known pipeline pigs. Because there is little leaked magnetic field in the aluminum material, the MFL technique cannot be used for the aluminum tube [8]. In the same way, the conventional magnetic particle (MP) testing method is inapplicable for the aluminum tube. The penetrant testing (PT) is suitable for the surface SCC and FCC. But these coatings and attachments on the aluminum tube should be cleaned deeply, which is time-consuming and difficult in the on-site environment. Besides, the PT is hard to detect the inner-wall crack. The radiographic testing (RT) can identify both surface and subsurface cracks clearly. However, the RT is harmful to human body, which is forbidden in many industrial fields. The ultrasonic testing (UT) is not an effective method for thin surface defects, especially the inner cracks. What's more, the UT needs coupling agent along with the probe. It is also a challenge to identify both inner and outer cracks using the UT in the aluminum tube [9].

Because of the non-ferromagnetic and high-conductive of the aluminum material, the current field perturbation NDT technique are excellent methods for the detection of surface cracks [10–12]. The conventional eddy current testing (ET), alternating current field measurement (ACFM), alternating current potential drop (ACPD) technique methods all belong to the current field perturbation NDT technique. When the surface cracks are present in the aluminum tube, the current field will turns obviously around the surface crack. Especially, the excitation frequency can be composited as multiple signals to penetrate different depth of the conductive material [13]. Bernieri, et al. proposed the multi-frequency ET for the detection and characterization of defects on conductive materials [14]. The frequency of the excitation signal is from direct current (DC) to dozens of MHz [15, 16]. He, Gao, Tian, Li, et al. combined the eddy current field and the thermography technique to identify sub-surface cracks [17–20]. Ge, et al., given the optimal time-domain feature for detection of non-surface crack using the pulsed ACFM technique [21]. Fan, Tian, et al., proposed the pulsed ET method to test subsurface defects and the wall thickness [22–24]. The pulse excitation signal gives more information about defects in frequency domain and time domain. However, these methods cannot distinguish and evaluate both inner and outer cracks effectively on the whole 360° cambered surface in a one pass scan. In our previous work, we proposed the CCFT system to measure outer cracks in pipeline using the external coaxial excitation coil [25]. As the high excitation frequency, the induced current gathers in the thin outer surface of the pipeline, which cannot penetrate the thick wall of the pipe. As a result, the inner-wall cracks cannot be inspected by the high frequency CCFT system. What's more, due to the corner joints and flanges, the external coaxial excitation coil is not suitable for the inspection of aluminum tubes in service time. The bobbin-type coil ET is a good ways to pass through the aluminum tube for the measurement of the inner defects in many piping systems [26]. In the industrial field, detection and evaluation both inner and outer cracks are heavy works in the long and large number of aluminum tubes [27]. It is significant to inspect all inner and outer surface cracks in the aluminum tube in real time by one scan.

In this paper, all inner and outer cracks in the aluminum tube are inspected by the double frequency CCFT method in a one pass scan. The circumferential current field is induced by a coaxial bobbin coil excited by the double frequency excitation signal which is composed of a low frequency component and a high frequency component. The induced circumferential current field can penetrate the wall of the aluminum tube due to the low frequency component. Meanwhile, it gathers in a skin layer in the inner wall due to the high frequency component. When the crack is present in the inner or outer surface, the circumferential current field turns around at the ends and the bottom of the crack resulting in the distorted space magnetic field which is measured by the multiple magnetic sensor arrays. All inner and outer cracks in the 360° cambered surface can be inspected by the double frequency CCFT method in a one pass scan.

The rest of the paper is organized as follows. Firstly, the FEM model of the double frequency CCFT is present to analyze the distorted electromagnetic field around the crack in Sect. 2. The probe and system of CCFT are developed in Sect. 3. The inner and outer cracks are identified, distinguished and evaluated by the double frequency CCFT system in Sect. 4. Finally, the conclusion and future work are drawn in Sect. 5.

2   Finite Element Method Model

2.1   Simulation Model

As shown in Fig. 1, the FEM model of CCFT probe is set up. In the FEM model, the bobbin coil is set as the excitation module which is coaxial with the aluminum tube. Two longitudinal cracks are produced axially (X direction) on the inner and outer surface of the aluminum tube. The depth of the longitudinal crack is in the radial direction (Y direction). The dimensions of the bobbin coil, aluminum tube and cracks are given in Table 1.

The bobbin coil is made up of 300 turns copper wires. The excitation signal is loaded on the bobbin coil with the frequency f. In the FEM model, the low and high frequency excitation signals are proposed to inspect the outer and inner cracks respectively. The bobbin coil induces the uniform circumferential current field in the aluminum tube. The penetration depth of the induced circumferential current field is given by skin effect Eq. (1).

$$\delta = \left(1\big/\pi u_{\mathrm{r}}u_0\gamma f\right)^{1/2} \tag{1}$$

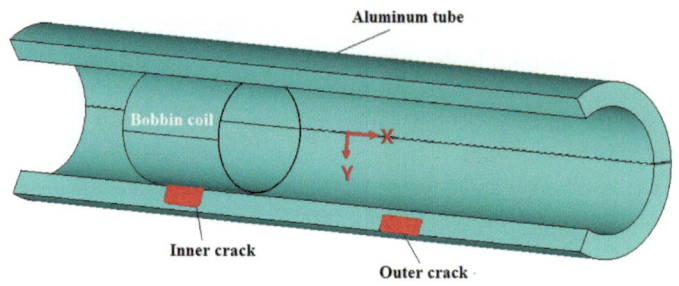

Fig. 1 FEM model of CCFT (Cutaway view)

Table 1 Dimensions of EFM model

Model	Diameter/mm	Length/mm	Width/mm	Depth/mm
Tube (D/d)	65/47	300	—	—
Bobbin coil	43	50	—	—
Outer crack	—	30	0.5	6
Inner crack	—	30	0.5	6

Table 2 Parameters of EFM model

f/Hz	Excitation current (mA)	u_r	u_0	γ	Skin depth (mm)
80	50	1	4πe−7	0.377e8	9.17
1000	50	1	4πe−7	0.377e8	2.59

where δ is the skin depth, u_r is the relative magnetic permeability, u_0 is the vacuum permeability, γ is the conductivity of the aluminium tube.

The parameters of the model are given in Table 2. It has been proven that 2–3 mm is the optimized skin depth to test the surface crack in the aluminum specimen. Thus the 1000 Hz is set as the high frequency component of the excitation signal [28]. To penetrate the wall thickness (9 mm) of the aluminum tube, the 80 Hz is set as the low frequency component of the excitation signal.

2.2   High Frequency Excitation Signal

In the FEM model, the frequency of the excitation signal is 1000 Hz and the amplitude of the excitation current field is 50 mA. Due to the skin effect, the penetration depth of the induced current field is 2.59 mm. Thus induced current field cannot be affected by the outer crack. As shown in Fig. 2a, the vector field of the induced circumferential current is extracted around the inner crack in the aluminum tube. When the current field is far from the inner crack, the current vector is uniform in the circumferential direction. Due to the material discontinuity of the inner crack, the circumferential current field is perpendicular to the longitudinal crack. The circumferential current turns clockwise at one end of the inner crack and turns counterclockwise at the other end of the inner crack. The current density around the inner crack is shown in Fig. 2b. There are two obvious peaks at the ends of the inner crack. Meanwhile, the current density decreases in the depth direction of the inner crack.

The perturbation current field makes the space magnetic field distorted. The decrease of the current density produces a trough in the axial magnetic field (Bx).

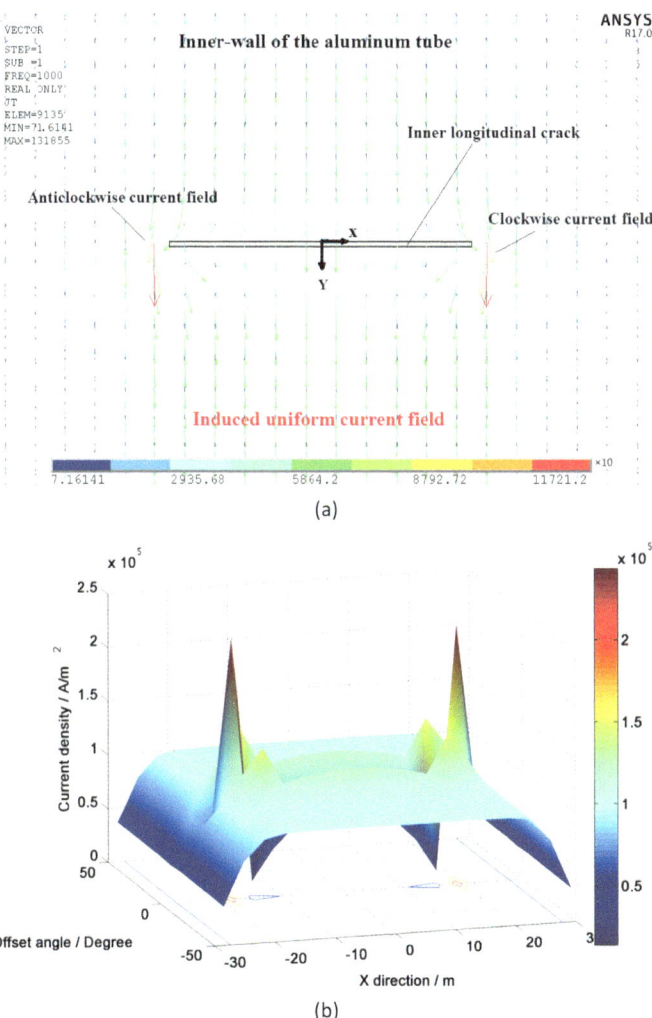

Fig. 2 Perturbation of the induced circumferential current field caused by the inner crack under the excitation frequency of 1000 Hz. **a** Vector current field (Internal view inside the aluminium tube). **b** Current density (Offset angle means the degree in the circumferential direction)

The opposite deflecting direction of the current field produces positive and negative peaks in the radial magnetic field (Bz). In practice, the lift-off of the bobbin coil (distance between the bobbin coil and the inner-wall of the aluminum tube) should be small to improve the current density. Thus the magnetic sensors can be set in the bobbin coil to pick up the distorted space magnetic field.

According to the Lenz's Law, the distortion of outer magnetic field is opposite to the magnetic field in the bobbin coil. Thus the distorted magnetic field around the inner crack can be measured by magnetic sensors in the bobbin coil. The space magnetic field is extracted below the bobbin coil at the lift-off of 5 mm (the distance between the inner wall and the magnetic sensor is 5 mm in radial direction). As shown in Fig. 3a, the Bx is normalized by the detection sensitivity Eq. (2):

$$SBx = (Bx - Bx_0)\big/Bx_0 \tag{2}$$

where S_{Bx} is the detection sensitivity of the Bx. Bx_0 is the background signal when the crack is not present.

As shown in Fig. 3a, the S_{Bx} shows a peak at the center of the inner crack. The S_{Bx} is mainly caused by the decrease of current field in depth direction of the inner crack. Thus the S_{Bx} reflects the crack depth. While, the Bz show positive and negative peaks at the tips of the inner crack, as shown in Fig. 3b. By measuring the distance of peaks (D_p) in the Bz, the length of the inner crack can be calculated.

2.3   Low Frequency Excitation Signal

The excitation frequency of the excitation signal is changed to 80 Hz, while keep the current amplitude same (50 mA). The perturbation characteristic of the circumferential current field around the inner crack is the same with that under the excitation frequency of 1000 Hz. But the current density, S_{Bx} and Bz are much weak under 80 Hz than that of 1000 Hz. Here we focus on the perturbation of current field caused by the outer crack. As shown in Fig. 4a, the induced circumferential current field penetrates the wall thickness because the penetration depth is 9.17 mm in aluminum tube at the excitation frequency of 80 Hz. The vector current field deflects around the bottom of the outer crack. As shown in Fig. 4b, the vector current field turns around with different deflecting direction at the end of the outer crack inside the aluminum tube. The current density below the outer crack is extracted in the inner wall of the aluminum tube, as shown in Fig. 4c. The current density shows a deep trough in the axial direction under the outer crack.

Comparing Fig. 4c with Fig. 2b, the current density caused by the outer crack under the excitation frequency of 80 Hz is much weaker than that caused by the inner crack under the excitation frequency of 1000 Hz. This is due to the induced current field gathers in the skin layer of the inner-wall at 1000 Hz frequency. Thus the 1000 Hz excitation frequency is suitable for inspection of inner crack. What's more, the induced current field decays exponentially in the aluminum tube to reach

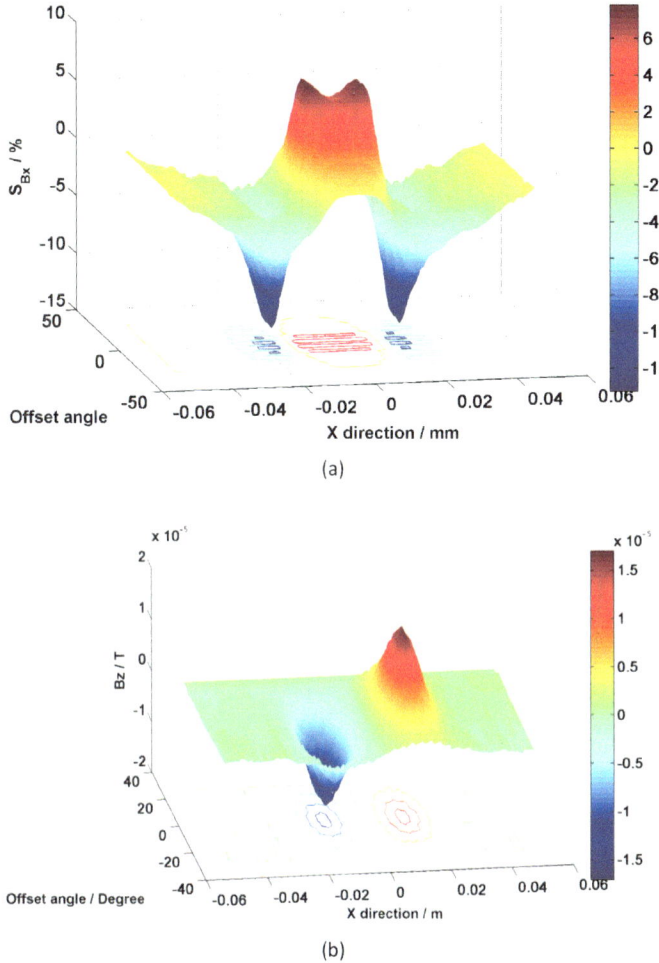

(a)

(b)

Fig. 3 Distorted space magnetic field in the bobbin coil under the excitation frequency of 1000 Hz. **a** Sensitivity of Bx (S_{Bx}). **b** Bz

the bottom of the outer crack at the excitation frequency of 80 Hz. Thus the distorted space magnetic field cause by the perturbation of current field is weak than that at the excitation frequency of 1000 Hz. As shown in Fig. 5a, the Bx is normalized by the Eq. (2). There is a weak peak in the S_{Bx} at the center of the outer crack, while there are two opposite peaks at the end of the outer crack, as shown in Fig. 5b.

Comparing Fig. 3a with Fig. 5a, the amplitude of the S_{Bx} caused by the outer crack under the excitation frequency of 80 Hz is much less than that caused by the inner crack under the excitation frequency of 1000 Hz. For the outer crack uder the excitation of 80 Hz, the S_{Bx} is mainly caused by the decay depth where the bottom of the outer crack is present. The Bz is mainly affected by the different deflecting

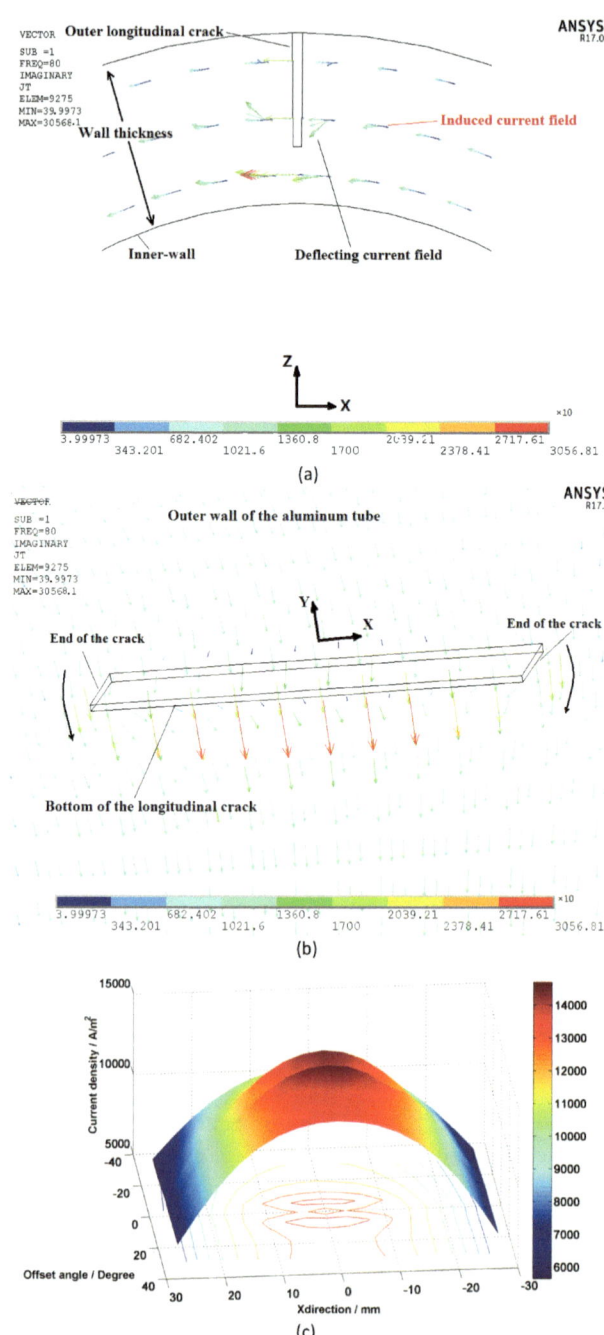

Fig. 4 Perturbation of current field caused by the outer crack under the excitation frequency of 80 Hz. **a** Vector current field flows at the bottom of the crack (Cross-sectional view). **b** Vector current field turns around at the ends of the crack (Perspective drawing from the outside). **c** Current density

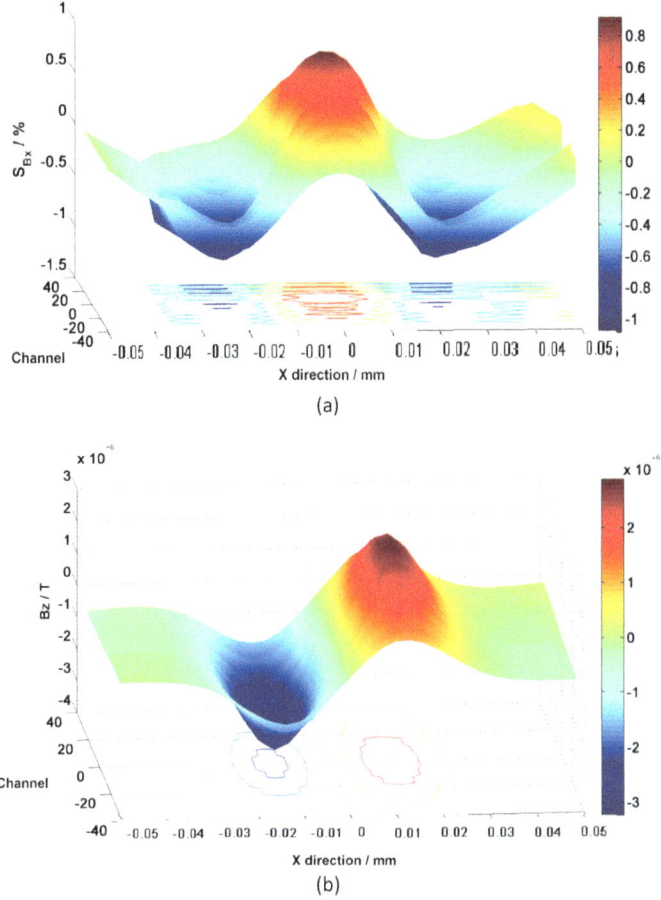

Fig. 5 Distorted magnetic field below the outer crack in the bobbin coil under the excitation frequency of 80 Hz. **a** S_{Bx}. **b** Bz

direction of the current field at the end of the outer crack. Similarly, the S_{Bx} and the D_p reflects the residual wall thickness (RWT) and the length of outer cracks respectively.

3 Testing System

3.1 Probe with Sensor Arrays

The bobbin-type probe is shown in Fig. 6a. The bobbin coil is made up of 300 turns enameled wires whose enameled wire is 0.15 mm. The bobbin coil is installed on a plastic yoke. The diameter of the bobbin coil is 43 mm and the length is 50 mm. As shown in Fig. 6b, there are two tunnel magneto resistance (TMR) chips on one printed circuit board (PCB) for the measurement of Bx and Bz respectively at one location. Several PCBs are installed in the plastic yoke, which is covered by the bobbin coil. According to the detection range of the TMR sensor inside the bobbin coil, 18 PCBs should be employed to cover the 360° inner and outer surface of the aluminum tube. To simplify the testing system, 5 PCBs (Including 5 TMR sensors to measure the Bx and another 5 TMR sensors to measure the Bz) are used to pick up Bx and Bz in the bobbin coil, as shown in Fig. 6c. The lift-off of TMR sensors are 5 mm in the radial direction as the same in the FEM model. The measured magnetic signals are amplified (The Bx is amplified 50 times and the Bz is amplified 100 times) preliminarily by the AD620 chip on the PCB.

3.2 Testing System

As shown in Fig. 7a, the double frequency CCFT system is developed. One signal generator produces a sinusoidal signal with the frequency of 1000 Hz and the other signal generator produces a sinusoidal signal with the frequency of 80 Hz. The two sinusoidal signals are added together by a summator which is made by the OPA134 chip [29]. The synthetic excitation signal is amplified by a power amplifier to keep the current amplitude 50 mA. The bobbin coil is excited by the double frequency synthetic excitation signal. Due to the double frequency components, the induced circumferential current field can penetrate the wall thickness and gathers in the inner thin layer of the aluminum tube at the same time. Thus the circumferential current field can turns around at both inner and outer cracks in the aluminum tube using the bobbin coil with double frequency excitation signals. The TMR magnetic sensor arrays measure the Bx and Bz caused by the perturbation of current field.

The response signals from the TMR sensors (Bx and Bz) and their unilateral frequency spectrum are shown in Fig. 8. As shown in the response signal, there are two frequency components in the time domain. As shown in the unilateral frequency spectrum, there are two peaks at the frequency of 1000 Hz and 80 Hz. It shows that the two frequency components are measured by TMR sensors in the response signals.

The two excitation signals, response signals are captured by the acquisition card and then transferred to the personal computer (PC). In the computer, the response signals are processed by a software developed by the LabVIEW and MATLAB. Firstly, the high frequency component of the response signal is extracted by the

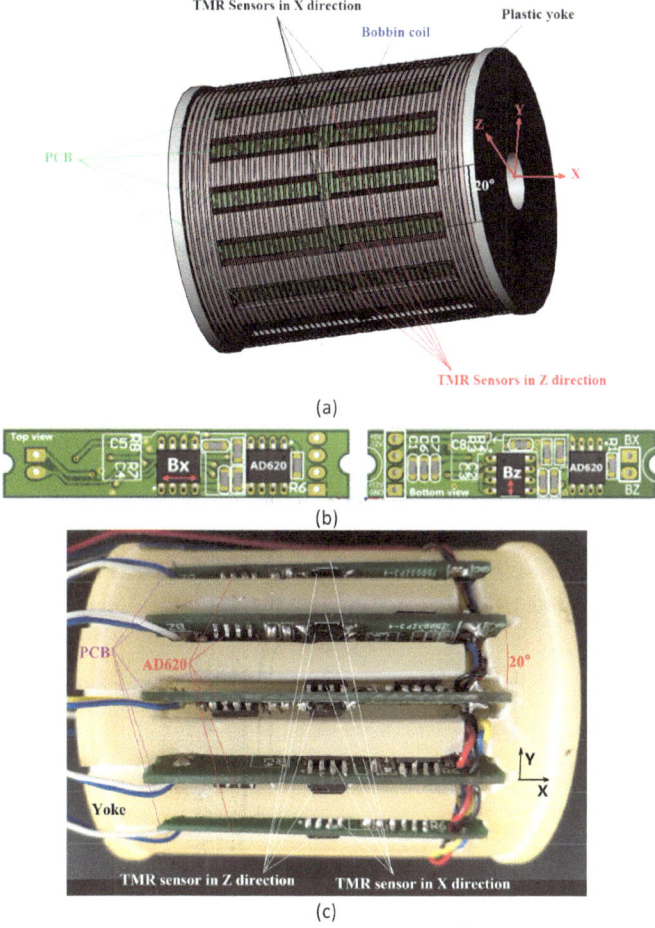

Fig. 6 Bobbin type probe with sensor arrays. **a** Structure of probe. **b** TMR sensors on PCB. **c** Five PCBs inside the plastic yoke

lock-in amplifier module using the 1000 Hz excitation signal. Meanwhile, the low frequency component of the response signal is extracted by the lock-in amplifier module using the 80 Hz excitation signal. Secondly, the low frequency component and high frequency component response signals are calculated by root-mean-square respectively. Thirdly, the Bx signals are normalized by the Eq. (2).

In the end, five S_{Bx} signals and five Bz signals with low frequency component are plotted respectively by the interpolation method. In the same way, the five S_{Bx} signals and five Bz signals with high frequency component are plotted respectively.

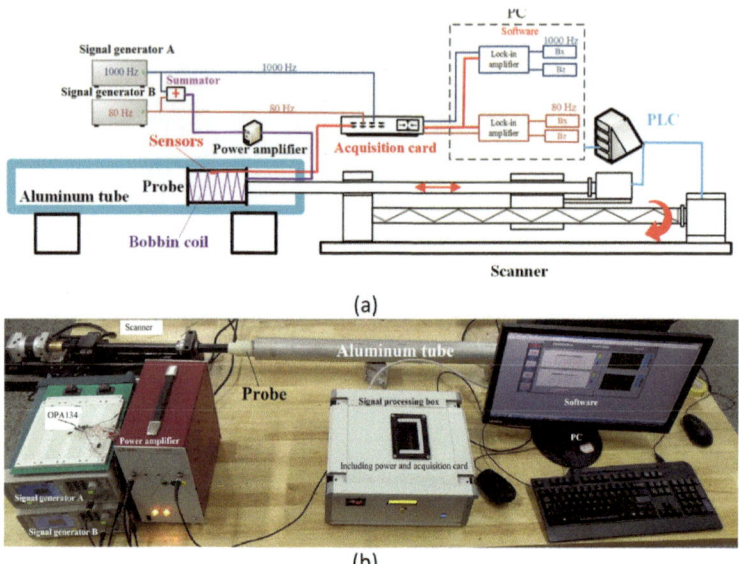

Fig. 7 Double frequency CCFT system. **a** Block diagram. b Photo of testing system

4 Inspection of Inner and Outer Cracks

4.1 Specimen

As shown in Fig. 9, there are four specimens with different size inner and outer longitudinal cracks. The specimens are aluminum tubes whose diameters are the same with the FEM model. Four inner and four outer cracks with different depths are introduced in the No. 1 and No. 2 specimens respectively, as shown in Fig. 9a, b. These cracks are with the same length (30 mm) and width (0.5 mm). Four inner and four outer cracks with different lengths are introduced in the No. 1 and No. 2 specimens respectively, as shown in Fig. 9c, d. These cracks are with the same depth (4 mm) and width (0.5 mm). The distance between the crack in each aluminum tube is 100 mm. The diameters of these cracks are shown in Fig. 9.

4.2 Inspection of Different Depth Cracks

The CCFT probe is pushed by a scanner to pass through inside the aluminum tube at the constant speed of 5 mm/s. The bobbin coil is excited by the double frequency synthetic excitation signal. Thus there are two frequency components in the induced circumferential current field. The high frequency component gathers in the thin layer

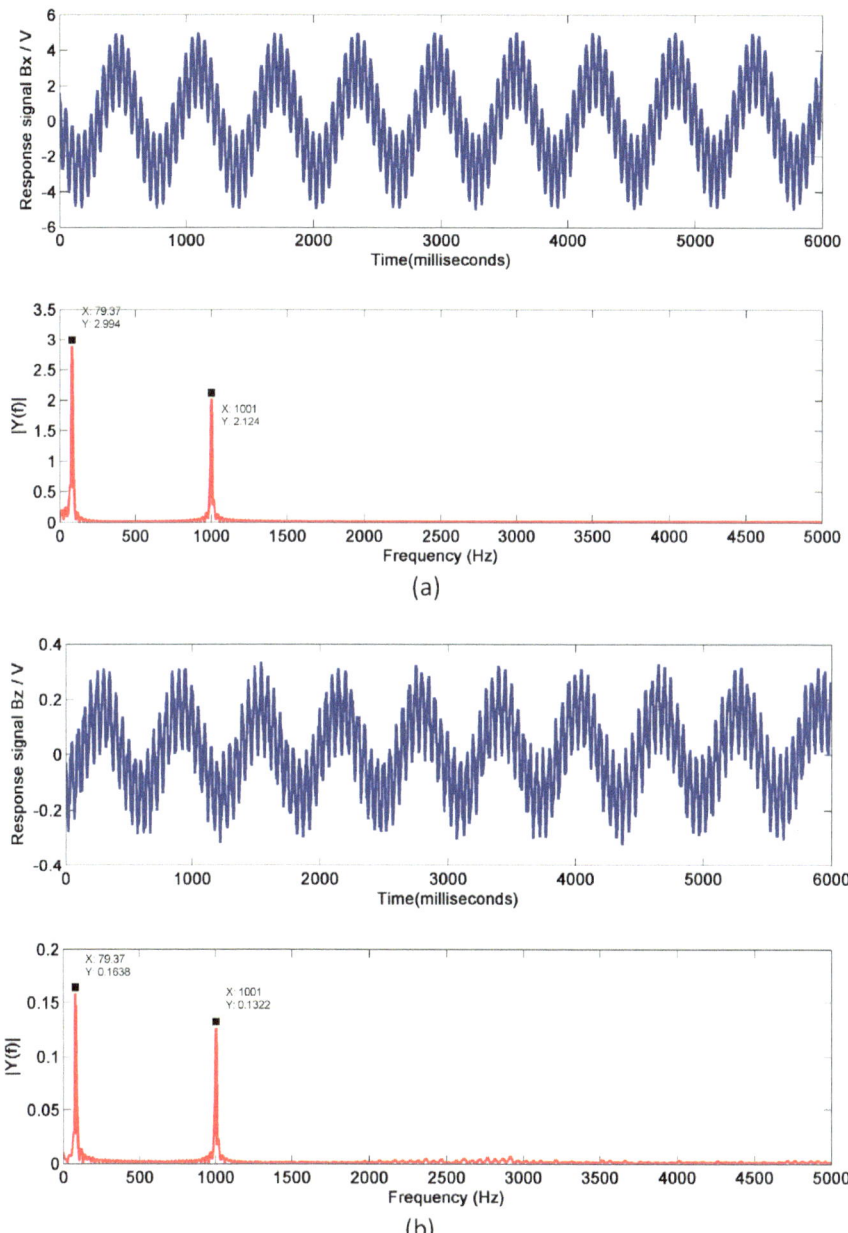

Fig. 8 Response signals from TMR sensors and unilateral frequency spectrum. **a** *Bx*. **b** *Bz*.

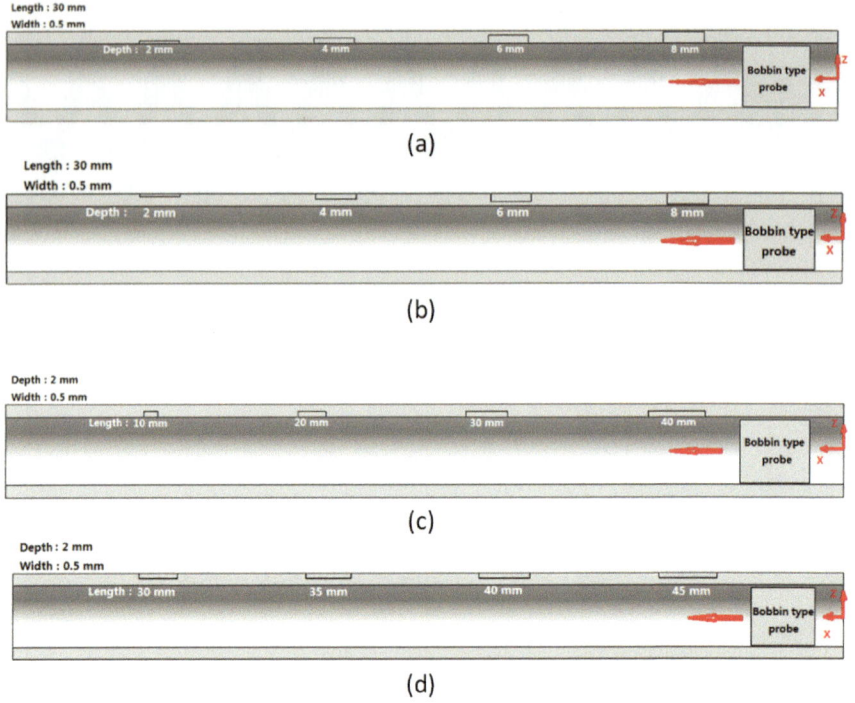

Fig. 9 Sizes of inner and outer longitudinal cracks in specimens. **a** No. 1 specimen with different depth inner cracks. **b** No. 2 specimen with different depth outer cracks. **c** No. 3 specimen with different length inner longitudinal cracks. **d** No. 4 specimen with different length outer longitudinal cracks

inside the aluminum tube. The low frequency component penetrates the wall thickness of the aluminum tube. The perturbation of current field will happen near both inner and outer cracks. The distortion of the space magnetic field can be measured by TMR sensor arrays. The distortion of space magnetic field caused by different depth inner cracks in No. 1 specimen is shown in Fig. 10.

As shown in Fig. 10a, c, there are four peaks in the S_{Bx} at 1000 and 80 Hz frequency components. Meanwhile, the Bz shows four peaks and troughs at the two frequency components at the same location, as shown in Fig. 10b, d. Comparing Fig. 10a with Fig. 10c, the distortion of magnetic caused by different depth inner cracks at the 1000 Hz frequency component is greater than that at the 80 Hz frequency component. The peak value of the S_{Bx} raises up with the growth of crack depth. Similarly, the peak of the Bz caused by each inner crack at the 1000 Hz frequency component is stronger than that at the 80 Hz frequency component, as shown in Fig. 10b, d. The peak of the Bz raises up as the crack depth increases. The D_p keeps the same because of the same length of cracks. The results show that the inner crack can be identified by both the 1000 Hz and 80 Hz frequency components. The distortion of space magnetic field caused by the inner crack at the 1000 Hz frequency component are much stronger

Fig. 10 Distortion of space magnetic field caused by different depth inner cracks. **a** S_{Bx} under at 1000 Hz frequency component. **b** Bz at the 1000 Hz frequency component. **c** S_{Bx} at the 80 Hz frequency component. **d** Bz at the 80 Hz frequency component

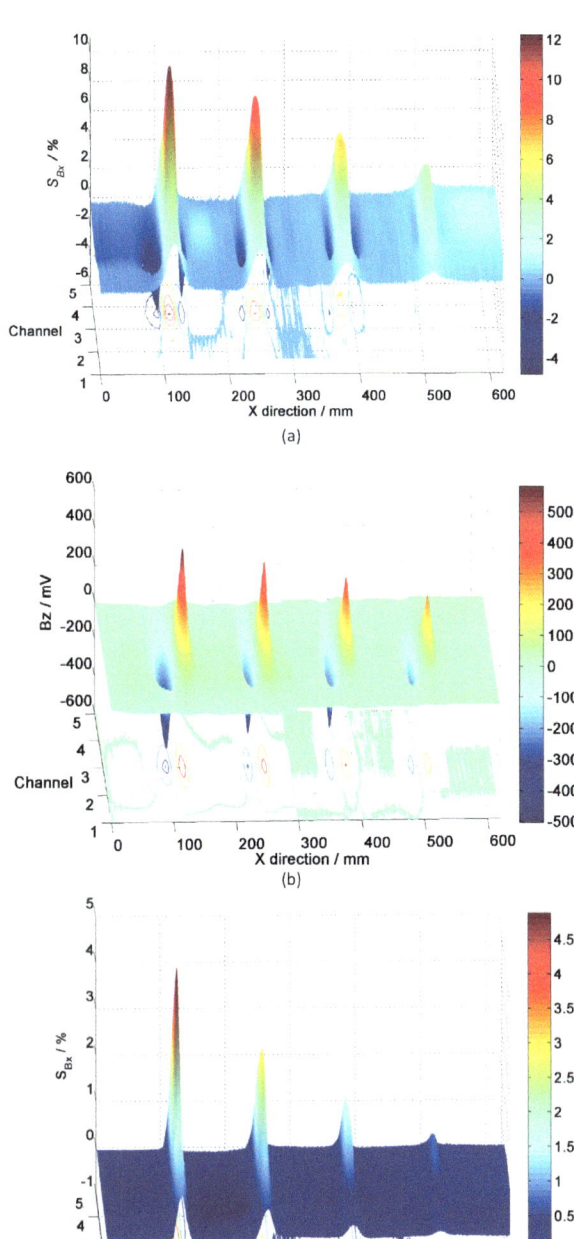

Fig. 10 (continued)

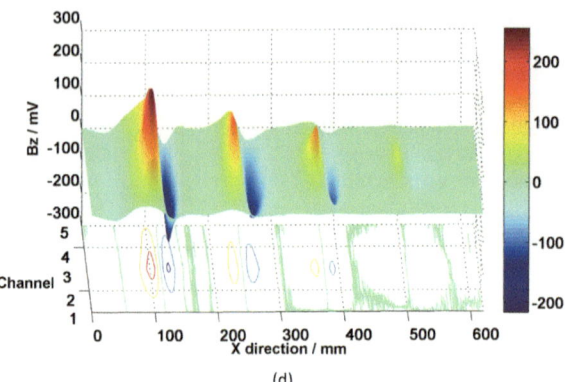

(d)

than that at the 80 Hz frequency component. The conclusions fit with the results from the FEM model. Thus S_{Bx} is a key parameter of inner crack depths, while D_p is a key parameter of inner crack lengths.

Due to the stronger distortion of magnetic field caused by inner cracks at the 1000 Hz frequency component, the S_{Bx} in Channel 3 from the Fig. 10a is used to evaluate the depth of the inner crack. The peaks of the S_{Bx} (S_p) are given in Table3. The first two cracks are set as the calibration cracks to evaluate the last two cracks by Eq. (3).

$$ED = 66.67Sp - 0.33 \tag{3}$$

where the S_p is the peak value of the S_{Bx}, the E_D is the evaluated depth of inner crack.

The E_D of the last two cracks is shown in Table 3. The absolute errors of the evaluated depth of the last two inner cracks are 0.2 mm and 0.3 mm. Because it is not a linear relationship between the crack depth and the S_p, the evaluated errors will increase for deeper inner cracks. However, when the crack depth varies in a certain rang (Due to the limited wall thickness), the evaluated depth is accurate. The measured errors are negligible in the industrial field. Thus the depth of inner crack is identified and evaluated by the 1000 Hz frequency component.

The No. 2 specimen is tested by the CCFT probe with the same parameters. As shown in Fig. 11, the distortion of magnetic field caused by different depth outer cracks are imaged. Due to the limited penetration depth of the induced current field at the excitation frequency of 1000 Hz, the current field cannot flows at the bottom of the outer crack. As a result, there is no obvious distortion in the S_{Bx}. Because the penetration depth of the induced current field can reach the deepest outer crack at

Table 3 Peak of S_{Bx} with different depth inner cracks

Depth/mm	2	4	6	8
S_p/%	3.5	6.5	9.8	12.1
E_D/mm	–	–	6.2	7.7

1000 Hz frequency component, the current field turns around slightly at the tip of the outer crack. Thus the B_z shows weak peaks at the deepest crack and no obvious distortion in other depth outer cracks.

Because the induced current field penetrates the wall of the aluminum tube under the excitation frequency of 80 Hz, the current field will flows at the bottom of the crack and turns around at the tip of the outer crack. Thus there are four obvious peaks in the S_{Bx} at the 80 Hz frequency component. Meanwhile, the B_z shows four peaks and troughs at the tips of outer cracks. The peak of the S_{Bx} increases as the residual wall thickness decreases, as shown in Fig. 11c. Comparing Fig. 11c with Fig. 10a, although the amplitudes of the distortion of magnetic field caused by outer cracks at 80 Hz frequency component are much weaker than that of inner cracks at 1000 Hz frequency component, the signal to noise ratio (SNR) is acceptable in the Fig. 11c, even for the shallow outer crack (Wall thickness is 7 mm). The D_p keeps the same because of the same length of outer cracks. It shows that outer cracks can be identified effectively by the 80 Hz frequency component, but cannot be identified by the 1000 Hz frequency component.

Similarly, the S_{Bx} in Channel 3 from the Fig. 11c is extracted to evaluate the residual wall thickness (RWT) of the aluminum tube. The S_p of the outer crack with the RWT is shown in Table 4.

The first two outer cracks are set as the calibration cracks. The residual wall thickness of last two cracks can be evaluated by Eq. (4).

$$ER = -1667Sp + 15.17 \qquad (4)$$

where the E_R is the evaluated residual wall thickness of the aluminum tube.

The evaluated residual wall thickness of the 6 mm depth outer crack is 3.0 mm, which equals the actual residual wall thickness. However, the evaluated residual wall thickness of the 8 mm depth outer crack is -12.2 mm. It does not accord with the actual residual wall thickness. This is due to the induced current field at the 1000 Hz frequency component has penetrated the residual wall thickness of the deepest outer crack. Thus the distorted current field and magnetic field are different from the first three outer cracks. The S_p of the last crack is much greater than that of the first three outer cracks. Meanwhile, this is a good phenomenon for the evaluation of the aluminum tube with little residual wall thickness, which has a higher risk of leakage. The last outer crack can be set as the trough crack in advance.

In conclusion, the outer crack can only be identified by the 80 Hz frequency component and the inner crack can be identified by the 1000 Hz and 80 Hz frequency components. Thus the outer and inner cracks can be distinguished by the response signals with low and high frequency components. The inner crack depth and the RWT of the outer crack can be evaluated by the peak of the S_{Bx} at the 1000 Hz and 80 Hz frequency component respectively. Thus the development of the crack depth in the aluminum tube can be evaluated by periodic detection using the double frequency CCFT method.

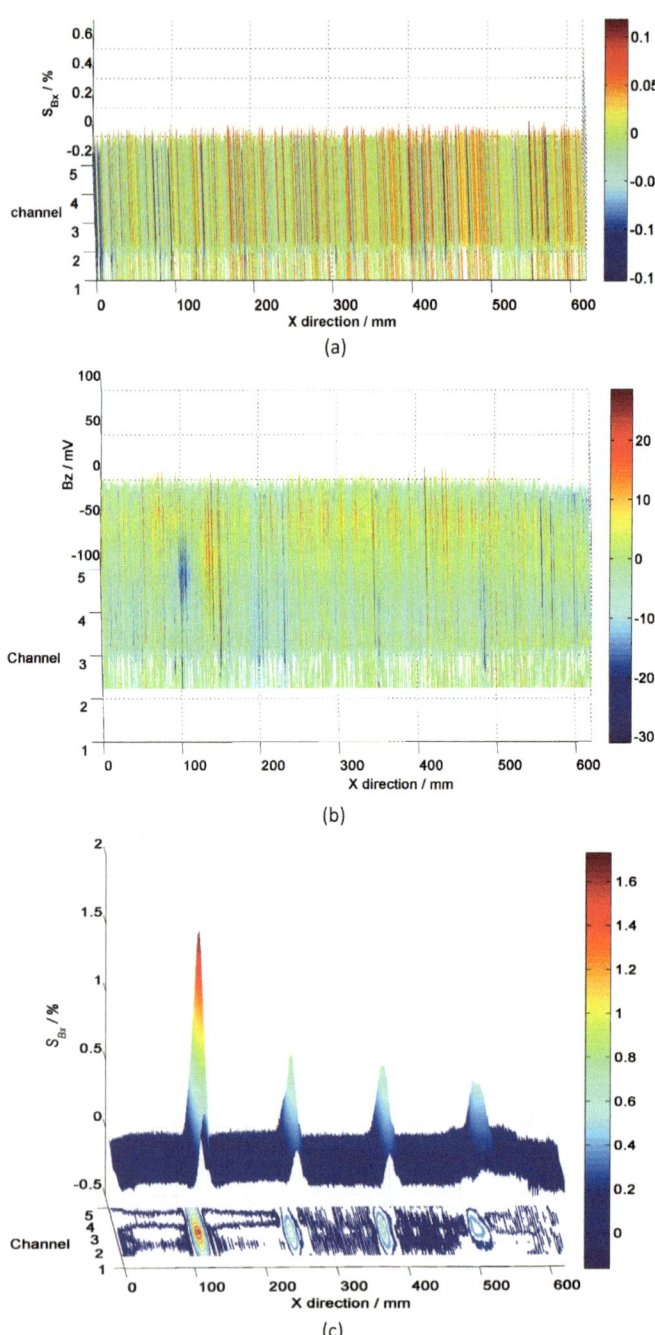

Fig. 11 Distortion of magnetic field caused by different depth outer cracks. **a** S_{Bx} at the 1000 Hz frequency component. **b** Bz at the 1000 Hz frequency component. **c** S_{Bx} at the 80 Hz frequency component. **d** Bz at the 80 Hz frequency component

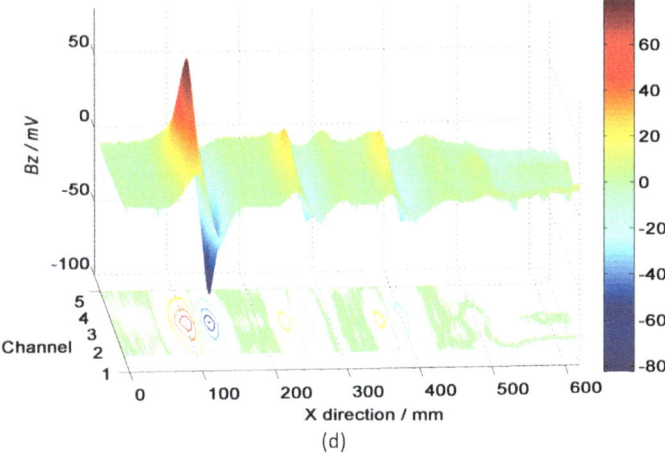

(d)

Fig. 11 (continued)

Table 4 S_p with different residual wall thickness

RWT/mm	7	5	3	1
$S_p/\%$	0.49	0.61	0.73	1.64
E_R/mm	–	–	3.0	Trough crack

4.3   Inspection of Different Length Cracks

The No. 3 specimen is tested by the CCFT probe at the same speed. Because response signals of inner crack are sensitive at the 1000 Hz frequency component, the response signals of different length inner cracks at 1000 Hz frequency component are presented only, as shown in Fig. 12.

The S_{Bx} shows four obvious peaks in the center of the inner crack, while the Bz shows four peaks and troughs at the tips of the inner crack. The four different length inner cracks are identified by the 1000 Hz frequency component. The D_p of the Bz with different length inner cracks is shown in Table 5.

The D_p is less than the length of the inner crack and the maximum absolute error is 2.8 mm. As a matter of fact, the absolute errors can be eliminated by calibration method. The first two cracks are set as the calibration crack. The length of the last two inner cracks can be estimated by Eq. (5).

$$EL = 1.04Dp + 1.04 \tag{5}$$

The estimated length (E_L) of the last two cracks is 29.54 mm and 39.73 mm respectively. The absolute errors are 0.46 mm and 0.27 mm. which is negligible in the industrial field.

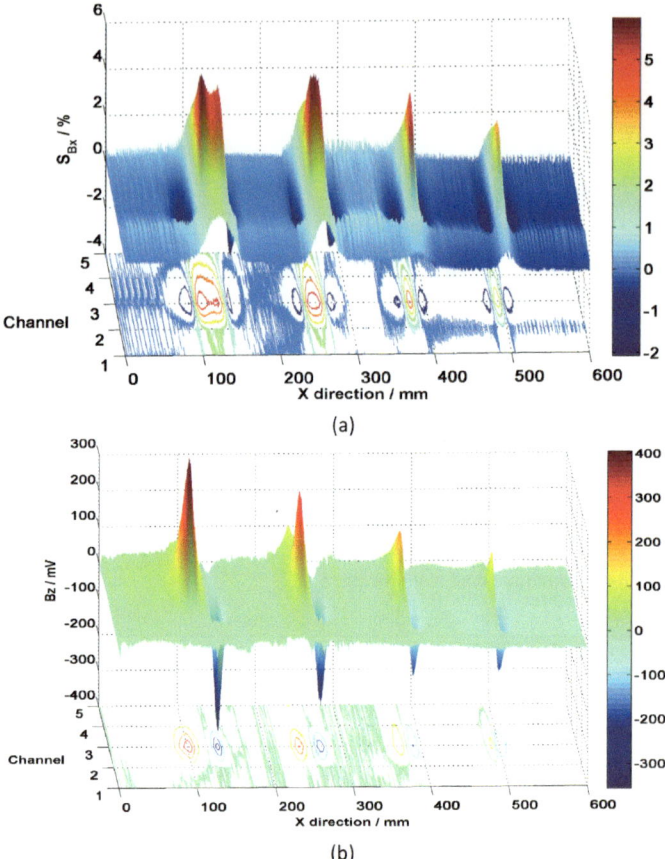

Fig. 12 Distortion of magnetic field caused by different length inner cracks at the 1000 Hz frequency component. **a** S_{Bx}. **b** Bz

Table 5 D_p with different length inner cracks

Length/mm	10	20	30	40
D_p/mm	8.6	18.2	27.4	37.2
E_L/mm	–	–	29.54	39.73

The No. 4 specimen is tested at the same speed. There are no obvious distorted response signals of the different length outer cracks at the 1000 Hz frequency component. The response signals at 80 Hz frequency component are shown in Fig. 13.

There are four obvious peaks at the center of the crack. Meanwhile there are four opposite peaks at the same location. The four different length outer cracks are identified by the 80 Hz frequency component. The D_p of the Bz is given in Table 6. The maximum absolute error is 3.8 mm. Similarly, the first two cracks are set as the

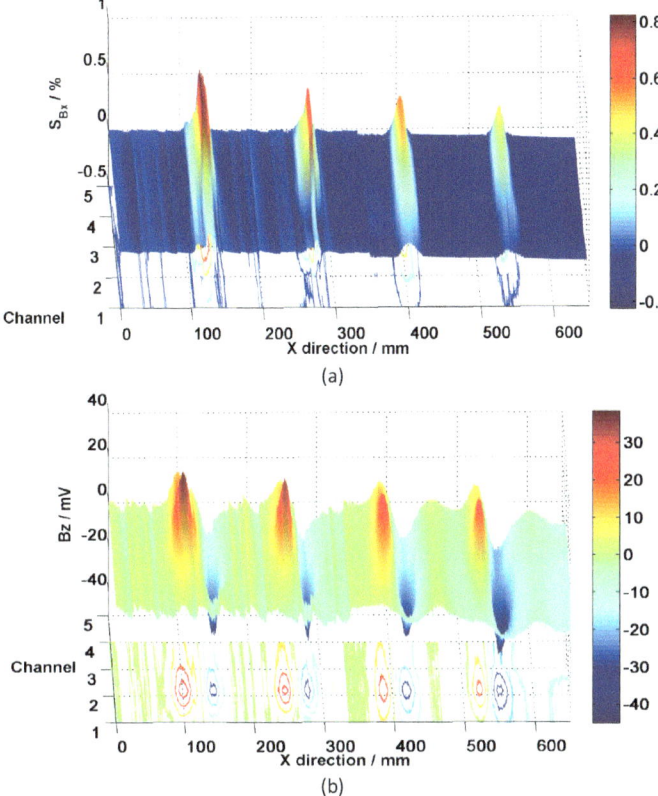

Fig. 13 Distortion of magnetic field caused by different length outer cracks at the 80 Hz frequency component. **a** S_{Bx}. **b** Bz

calibration crack. The length of the last outer cracks can be estimated by Eq. (6).

$$EL = 1.02Dp + 3.06 \qquad (6)$$

where E_L is the estimated length of the crack.

The E_L of the last two cracks is shown in Table 4. The absolute error is 0.02 mm and 0.29 mm.

In a word, the length of inner and outer cracks can be estimated by the D_p at the 1000 Hz and 80 Hz frequency component respectively. The extension of inner and

Table 6 D_p with different length outer cracks

Length/mm	30	35	40	45
D_p/mm	26.4	31.3	36.2	41.4
E_L/mm	–	–	39.98	45.29

outer cracks can be monitored by the periodic detection using the double frequency CCFT method.

5 Conclusions and Further Work

The double frequency CCFT is presented for the detection and evaluation of both inner and outer cracks in the aluminum tube in a one pass scan. The FEM model of the CCFT is set up to analyze the disturbed current field and distorted magnetic field around inner and outer longitudinal cracks with the excitation frequency of 1000 Hz and 80 Hz respectively. The CCFT probe is built with TMR sensor arrays inside the bobbin type excitation coil. The 1000 Hz frequency excitation signal and 80 Hz frequency excitation signal are added together as the synthetic excitation signal to excite the bobbin coil in the probe. The response signals with 1000 Hz and 80Hz frequency components are separated by the lock-in amplifier module in the software. The CCFT system is developed for the inspection of different size inner and outer longitudinal cracks in the aluminum tube. The results show that outer cracks can only be identified by the 80 Hz frequency component, while inner cracks can be identified by the 1000 Hz and 80 Hz frequency component. Thus the outer and inner cracks can be distinguished by the double frequency components. The RWT of outer cracks and inner crack depth can be evaluated by the S_{Bx}. The crack length can be measured by the D_p of the Bz. As a result, both the inner and outer cracks can be identified, distinguished and evaluated using the double frequency CCFT method with TMR sensor arrays inside the aluminum tube in a one pass scan. The double frequency CCFT probe and system can be used in the in-service inspection by crawling through the aluminum tube. Further work will focus on the inspection of samples with complex defects, the development of the portable instrument.

References

1. C.J. Earls, Bayesian inference of hidden corrosion in steel bridge connections: Non-contact and sparse contact approaches. Mech. Syst. Signal Process. **41**, 420–432 (2013)
2. Z. Li, H. Yang, X. Hu, J. Wei, Z. Han, Experimental study on the crush behavior and energy-absorption ability of circular magnesium thin-walled tubes and the comparison with aluminum tubes. Eng. Struct. **164**, 1–13 (2018)
3. Z. Wang, Z. Li, W. Zhou, D. Hui, On the influence of structural defects for honeycomb structure. Compos. Part B **142**, 183–192 (2018)
4. H.R. Hajibagheri, A. Heidari, R. Amini, An experimental investigation of the nature of longitudinal cracks in oil and gas transmission pipelines. J. Alloy. Compd. **741**, 1121–1129 (2017)
5. M. Javidi, S.B. Horeh, Investigating the mechanism of stress corrosion cracking in near-neutral and high pH environments for API 5L X52 steel. Corros. Sci. **80**, 213–220 (2014)

6. M. Win, A.R. Bushroa, M.A. Hassan, N.M. Hilman, A. Ide-Ektessabi, A contrast adjustment thresholding method for surface defect detection based on mesoscopy. IEEE Trans. Ind. Informat. **11**(3), 642–649 (2015)
7. Du-Ming Tsai, Shih-Chieh Wu, and Wei-Yao Chiu, Defect detection in solar modules using ICA basis images, IEEE Trans. Ind. Informat., 9(1)(2013)122–131.
8. Ju-Won Kim and Seunghee Park, Magnetic flux leakage sensing and artificial neural network pattern recognition-based automated damage detection and quantification for wire rope non-destructive evaluation, *Sensors*, 18(109)(2018).
9. H.R. Vanaei, A. Eslami, A. Egbewande, A review on pipeline corrosion, in-line inspection (ILI), and corrosion growth rate models. Int. J. Pres. Ves. Pip. **149**, 43–54 (2017)
10. M.B. Fan, Q. Wang, B.H. Cao, B. Ye, A.I. Sunny, G.Y. Tian, Frequency optimization for enhancement of surface defect classification using the eddy current technique. Sensors. **16**(5), 1–16 (2016)
11. X.A. Yuan, W. Li, G.M. Chen, X.K. Yin, J.H. Ge, Circumferential current field testing system with TMR sensor array for non-contact detection and estimation of cracks on power plant piping. Sensor. Actuat. A: Phys. **263**, 542–553 (2017)
12. S.S. Rao, M. Liu, F. Peng, B. Zhang, H.J. Zhao, Signal sensitivity of alternating current potential drop measurement for crack detection of conductive substrate with tunable coating materials through finite element modeling. Meas. Sci. Technol. **27**, 125004 (2016)
13. A. Bernieri, G. Betta, L. Ferrigno, M. Laracca, S. Mastrostefano, Multifrequency excitation and support vector machine regressor for ECT defect characterization. IEEE Trans. Instrum. Meas. **63**, 1272–1280 (2014)
14. Y. Gotoh, K. Sakurai, N. Takahashi, Electromagnetic inspection method of outer side defect on small and thick steel tube using both AC and DC magnetic fields. IEEE Trans. Magn. **45**, 4467–4470 (2009)
15. H. Saguy, D. Rittel, Application of ac tomography to crack identification. Appl. Phys. Lett. **91**, 084104 (2007)
16. Y.Z. He, R.Z. Yang, Eddy current volume heating thermography and phase analysis for imaging characterization of interface delamination in CFRP. IEEE Trans. Ind. Informat. **11**(6), 1287–1297 (2015)
17. B. Gao, W.L. Woo, G.Y. Tian, H. Zhang, Unsupervised diagnostic and monitoring of defects using waveguide imaging with adaptive sparse representation. IEEE Trans. Ind. Informat. **12**(1), 405–416 (2016)
18. K.J. Li, G.Y. Tian, L. Cheng, A.J. Yin, W.P. Cao, S. Crichton, State Detection of Bond Wires in IGBT Modules Using Eddy Current Pulsed Thermography. IEEE Trans. on Power Electron. **29**, 5000–5009 (2014)
19. Y.L. Gao, G.Y. Tian, Emissivity correction using spectrum correlation of infrared and visible images. Sensor. Actuat. A: Phys. **270**, 8–17 (2018)
20. J.H. Ge, G.M. Chen, W. Li et al., Investigation of optimal time-domain feature for non-surface defect detection through pulsed alternating current field measurement technique. Meas. Sci. Technol. **29**, 015601 (2017)
21. M.B. Fan, B.H. Cao, A.I. Sunny et al., Pulsed eddy current thickness measurement using phase features immune to liftoff effect. NDT & E Int. **86**, 123–131 (2017)
22. Y.Z. He, G.Y. Tian, M.C. Pan, D.X. Chen, Non-destructive testing of low-energy impact in CFRP laminates and interior defects in honeycomb sandwich using scanning pulsed eddy current. Compos. Part B **59**, 196–203 (2014)
23. A. Sophian, G. Y. Tian, and M. B. Fan, Pulsed Eddy Current Non-destructive Testing and Evaluation: A Review, Chin. J. Mech. Eng.-En.30(3)(2017)500–514.
24. X.A. Yuan, W. Li, G.M. Chen, X.K. Yin, W.C. Yang, J.H. Ge, Two-step interpolation algorithm for measurement of longitudinal cracks on pipe strings using circumferential current field testing system. IEEE Trans. Ind. Informat. **14**(2), 394–402 (2018)
25. J. Lee , J. Jun, J. Kim, H. Choi, and Minhhuy Le, Bobbin-type solid-state hall sensor array with high spatial resolution for cracks inspection in small-bore piping systems, IEEE Trans. Magn.48(2012)3704–3707.

26. Y.F. Zhao, J. Mehnen, A. Sirikham, R. Roy, A novel defect depth measurement method based on nonlinear system identification for pulsed thermographic inspection. Mech. Syst. Signal Process. **85**, 382–395 (2017)
27. X.A. Yuan, W. Li, G.M. Chen et al., Frequency optimization of circumferential current field testing system for highly sensitive detection of longitudinal cracks on pipe string. Insight **59**(7), 378–382 (2017)
28. F.J. Gan, W.Y. Li, J.B. Liao, New feature for evaluation of subsurface defects via multi-frequency alternating current field signature method. AIP Adv. **8**, 015026 (2018)
29. T. Reyno, P.R. Underhill, T.W. Krause, C. Marsden, D. Wowk, Surface profiling and core evaluation of aluminum honeycomb sandwich aircraft panels using multi-frequency eddy current testing. Sensors **17**, 2144 (2017)

Novel Phase Reversal Feature for Inspection of Cracks Using Multi-frequency Alternating Current Field Measurement Technique

Abstract Aluminum and its alloys have been widely used in aerospace and other industrial fields. Aluminum and its alloy structures are prone to surface and subsurface cracks when they are used in harsh environments. In this paper, a novel phase reversal feature is found to classify and evaluate cracks using the multi-frequency alternating current field measurement (ACFM) technique. The theoretical model of the phase reversal feature is developed. The distorted electromagnetic field and response signals of surface and subsurface cracks are analyzed by the finite element method. The multi-frequency ACFM testing system is set up. The experiments are carried out to test surface and subsurface cracks. The results show that the response signals of the surface and subsurface cracks have distinct characteristic due to the phase reversal feature using the multi-frequency ACFM technique. The surface and subsurface cracks can be classified by the amplitude reversal phenomenon of the Bz signal caused by the novel phase reversal feature. The buried depth of the subsurface crack can be evaluated accurately by the reversal frequency component.

Keywords Phase reversal feature · Surface and subsurface cracks · Multi-frequency ACFM · Classification and evaluation

1 Introduction

Aluminum is the second largest metal material next only to the steel material [1]. Because of the advantages of the light weight, the high strength, the good corrosion resistance, and the high thermal conductivity, aluminum and its alloys have been widely used in the power generation, the transportation, the construction and the aerospace industries [2–4]. The aluminum and alloy structures usually surfer from the huge wind load, the thermal shock, and other extreme conditions. As a result, cracks initiate and propagate in the surface and subsurface of the aluminum structure, which threatens the safety of the structure [5, 6]. Hence, it is of prime importance to inspect and evaluate these cracks for the health assessment of the structure.

© The Author(s) 2024
X. Yuan et al., *Recent Development of Alternating Current Field Measurement Combine with New Technology*, https://doi.org/10.1007/978-981-97-4224-0_5

It is well known that the electromagnetic nondestructive testing (ENDT) technique such as the eddy current testing (ET) is an excellent method for the detection of surface cracks in the aluminum. Generally, the excitation signal of the conventional ET is a high frequency sinusoidal signal [7]. Due to the skin effect, the conventional ET cannot be used to detect subsurface defects. The pulsed excitation signal and the multi-frequency excitation signal are introduced by some scholars [8–11].

In the pulsed eddy current testing (PECT) field, a strong and transient square pulsed excitation signal is loaded on the excitation coil. Because of the strong energy and the rich spectrum information, the response signal contains more characterizations about the subsurface defect. Ali Sophian et al. [12] developed a new PCA-based feature extraction method for the PECT. The method can effectively classify defects, and its performance is better than the conventional method using the response peak characteristics. He et al. [13] employed the peak amplitude and zero-crossing time of response signal in time domain to detect and characterize defects. Giguere, Fan, Tian, et al. [14–16] found the lift-off point of intersection (LOI) in the PECT method. The LOI is regarded as the potential feature for the evaluation of the subsurface defect. However, due to the wide frequency band in the frequency domain, the energy wasting is a great and unavoidable malpractice. Especially, the energy of the high frequency component is limited, which reduces the detection sensitivity of surface defect. Besides, because of the critical time domain analytical method, there are many features and factors that affect the testing results using the PECT.

In the multi-frequency eddy current testing (MFECT) field, two and more sine signals are added together as a multi-frequency excitation signal. The multi-frequency response signal can achieve specific permeation depth and obtain accurate characteristic information about the defect. Bernieri et al. [17] proposed a combination of a multi-frequency excitation and an optimized support vector machine for regression (SVR) for the reliable estimation of the geometrical characteristics of a thin defect. Zhang et al. [18] proposed a new approach to measure the multi-layer conductive coatings' thickness based on the MFECT. Xie et al. [19] presented a novel frequency-band-selecting pulsed eddy current testing (FSPECT) method for the detection of local wall thinning defects in a certain depth range. In their work, a specific excitation signal was designed to replace the pulsed signal, which improved the detection sensitivity of subsurface defects in a certain depth range significantly. Gao et al. [20] presented the spectrum method for the identification and classification of defects using the MFECT method. From the above, the multi-frequency excitation method has gained great progress in the ET field. However, there are still two critical issues should be addressed for the inspection of cracks in the aluminum material. Firstly, the ET method is sensitive to the lift-off effect. Thus, there are many interference signals when the probe variations above the roughened surface, such as the weld. Secondly, the MFECT usually is presented to detect the surface and subsurface cracks in the aluminum by the amplitude of the response signal. Both surface and subsurface cracks have response signals from the multi-frequency excitation. The classification and evaluation of surface and subsurface cracks is still a challenge only by the amplitude characteristic when the buried depth of subsurface cracks is shallow.

The alternating current field measurement (ACFM) is an emerging ENDT method for the detection and evaluation of cracks [21–24]. Due to the advantages of the non-contact testing, the high tolerance to lift-off and the quantitative evaluation ability, it has been widely used in the petrochemical industry [25–29], the electric power industry and the rail transportation industry. As the same with the conventional ET, the ACFM probe is excited by a sinusoidal signal. Thus, the ACFM is usually used to inspect surface cracks. In our previous work, a double frequency circumferential current field testing (CCFT) method was presented for the detection and evaluation of both inner and outer cracks in the aluminum tube [30]. The amplitude feature is used to classify and evaluate the surface and subsurface cracks. However, the amplitude of response signal is affected by the length, the depth or the buried depth of the crack. Thus, the evaluation result is obtained at the certain conditions, such as the same length crack, the large buried depth. Thus, new features should be find to classify and evaluate surface and subsurface cracks in the aluminum using the multi-frequency ACFM.

In this paper, a novel phase reversal feature is found to classify and evaluate surface and subsurface cracks based on the multi-frequency ACFM. The phase reversal feature is a inherent feature of the multi-frequency excitation ACFM, which is not affected by the dimensionality of the subsurface crack. Thus the surface and subsurface cracks can be classified and the buried depth can be evaluated accurately. The rest of the paper is organized as follows. In Sect. 2, the physical principle and the 3D finite element method (FEM) model of the multi-frequency ACFM are developed to analyze the phase reversal feature. In Sect. 3, the multi-frequency ACFM system is developed. In Sect. 4, the surface and subsurface cracks are detected. The surface and subsurface cracks are classified using the amplitude reversal phenomenon of the Bz signal. The buried depth of the subsurface crack is evaluated by the reversal frequency. In Sect. 5, the conclusion and further work are drawn.

2 Methodology

A. Physics Principle

In the classical ACFM model, an induced uniform current field is excited on the aluminum specimen when an alternating excitation signal is loaded on the excitation coil, as shown in Fig. 1a. Due to the skin effect, the induced currents mainly gather in the thin surface of the specimen. When a surface crack is presented, the surface current field will be disturbed. It makes the magnetic field distorted. Thus, the surface crack can be detected by measuring the distorted magnetic field.

Generally, we only focus on the surface thin induced current. In fact, the induced current field penetrates inside the aluminum specimen. The induced currents at different depths inside the aluminum specimen can be given by Eq. (1). It can be seen that the amplitude of the induced current field decreases exponentially and the phase of the induced current field lags linearly with the increase of depth.

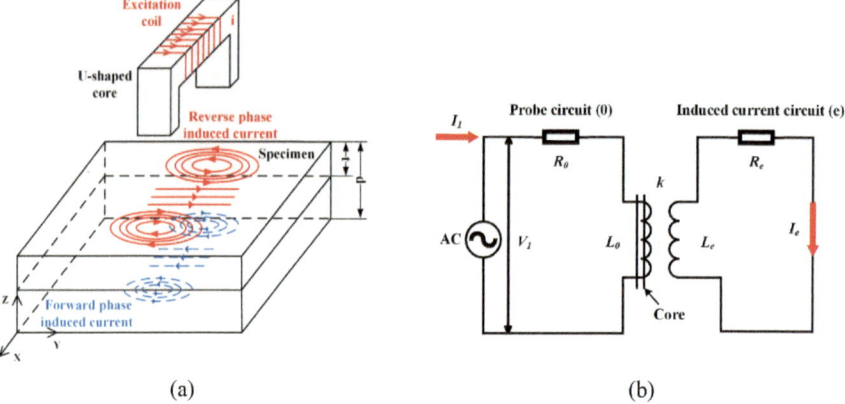

(a)   (b)

Fig. 1 Theoretical model. **a** ACFM model. **b** Transformer circuit model

$$J_Z = J_0 e^{-\frac{z}{\delta}(1+j)} \tag{1}$$

where J_z is the induced current at the depth Z, J_0 is the current in the surface of the specimen. When the current density in a certain depth of the specimen decays to 1/e of that on the surface of the specimen, the certain depth is called skin depth, as given in Eq. (2).

$$\delta = \left(1/\pi\rho f u_r u_0\right)^{1/2} \tag{2}$$

where δ is the skin depth, f is the excitation frequency, u_r is the relative magnetic permeability, u_0 is the vacuum permeability, ρ is the electric conductivity of the specimen.

The phase change of the induced current field can be modelled as a transformer circuit [31], as shown in Fig. 1b. The excitation coil is regarded as the input side of the transformer and the specimen is regarded as the output of the transformer. The electromagnetic coupling equivalent circuit follows the Kirchhoff's law, as given in Eqs. (3)–(5).

$$V_1 = I_1(R_0 + j\omega L_0) + I_e(j\omega M) \tag{3}$$

$$0 = I_e(R_e + j\omega L_e) + I_1(j\omega M) \tag{4}$$

$$M = k\sqrt{L_0 L_e} \tag{5}$$

where k is the coefficient of coupling between the two inductors. M is the mutual inductance in the circuits.

I_e is given in Eq. (6) by rearranging Eq. (4).

$$I_e = -I_1 \frac{j\omega M}{R_e + \omega^2 L_e} = -I_1 \frac{\omega M}{R_e^2 + \omega^2 L_e^2}(\omega L_e + jR_e) \tag{6}$$

For plane wave excitation, the resistive and reactive components of the induced current impedance are equal in magnitude ($R_e = \omega L_e = X_e$) [2]. So that I_e is calculated in Eq. (7).

$$I_e = -I_1 \omega L_0 \frac{k^2}{2}(1+j) \equiv I_1 \omega L_0 \frac{k^2}{\sqrt{2}} e^{-\frac{3\pi}{4}j} \tag{7}$$

Therefore, when the phase of the excitation current is $0°$, the phase of the induced current on the surface of the aluminum specimen is $-135°$. In addition, the phase of the induced current field lags linearly with the increase of penetration depth. The phase suddenly reverses from $-180°$ to $180°$ at a specific penetration depth, which is called a phase reversal feature in this paper. The phase reversal feature represents a change of the induced current direction. When the thickness of the aluminum specimen is infinite, the direction of the induced current changes periodically. In practice, the plate thickness is limited. The induced currents inside the specimen are more severely attenuated and the disturbance around cracks is weaker. Thus, only one time phase change of the induced current field is considered in this paper. So the induced currents in the specimen can be divided into the forward phase current and the reverse phase current.

In this paper, the phase of the excitation signal is set $0°$. The reverse phase induced current field is in the surface of the aluminum specimen. The forward phase induced current field is under the reverse phase induced current field in the depth direction of the specimen, as shown in Fig. 1a. The phase reversal feature can take place with variational depths as the excitation frequency is different.

For surface cracks, when the crack is shallow, only the reverse phase induced current field is disturbed, as shown in Fig. 2a. The induced currents turn around at the tips of the surface crack in one direction all the time. Thus, the space magnetic field distorted in the vertical direction (Called Bz) is always in one direction because the reverse phase induced current field deflects in the same direction. When the crack is deep, the reverse phase and forward phase induced current fields are disturbed simultaneously, as shown in Fig. 2b. The two induced current fields turn around at the tips of the surface crack at the same time and the deflection directions are opposite. However, the density of the reverse phase induced current field is much greater than that of the forward phase current field. Thus, the direction of the Bz signal is still determined by the deflection direction of the reverse phase induced current field. In a conclusion, the direction of the distorted magnetic field is always the same direction regardless of the depth of surface cracks.

For subsurface cracks, when the crack is only in the forward phase induced current field [32], as shown in Fig. 2c. The induced currents turn around at the tips of the subsurface crack in one direction and the direction of the Bz signal is always in one direction. However, when the subsurface crack is in the reverse phase and forward phase induced current fields, as shown in Fig. 2d, the turned direction of

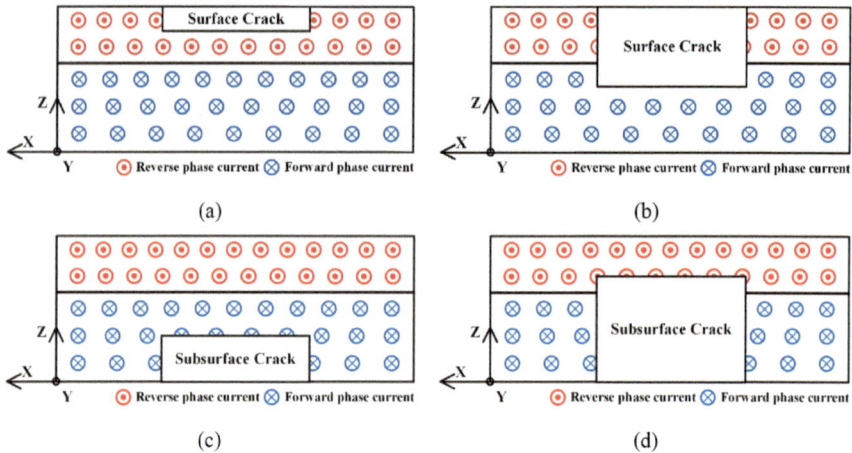

Fig. 2 Cracks in different current fields. **a** Surface crack in the reverse phase current field. **b** Surface crack in the reverse and forward phase current fields. **c** Subsurface crack in the forward phase current field. **d** Subsurface crack in the reverse and forward phase current fields

the reverse phase current filed is opposite to that of the forward phase current filed. What's more, the reverse phase current filed is stronger. As a result, the direction of the distorted magnetic field changes to the opposite direction due to the deeper buried depth of the surface cracks (Called amplitude reversal phenomenon). Thus, the phase reversal feature can be used to classify the surface crack and the subsurface crack when the multi-frequency excitation ACFM is carried out. When the phase reversal feature appears near the buried depth, there will be no obvious peak or trough (Called transitional signal) in the Bz signal. The buried depth can be obtained by the transitional signal using the multi-frequency response signal.

B. FEM Modeling and Analyzing

To verify the theoretical model proposed above, a 3D finite element method (FEM) model of ACFM was set up by the COMSOL software, as shown in Fig. 3. The simulation model consisted of a specimen, a U-shape core, a coil, and an air domain. The excitation coil (500 turns) was wound on the U-shape Mn–Zn ferrite yoke. The excitation signal was loaded on the excitation coil. The amplitude of the excitation current was 100 mA and the frequency was 1 kHz. The specimen was an aluminum plate, whose conductivity was 3.774×10^7 S/m and relative permeability was 1. The lift-off value of the probe was 1 mm.

The induced uniform current field at different depths in the aluminum were extracted, as shown in Fig. 4a. It is worth noting that the direction of the induced current field turns to the opposite direction at the depth of 3 mm. This is because the phase of induced current field changes from $-180°$ to $180°$ from depth 2 mm to 3 mm. The crossing $0°$ phase point is called phase reverse point (PRP) at this specific depth. The current density attenuates exponentially and the phase of the induced

Fig. 3 FEM model

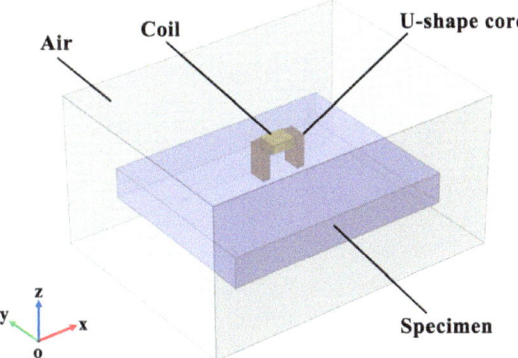

current lags linearly with the increasing depth of the aluminum specimen, as shown in Fig. 4b. This phenomenon is consistent with the theoretical model proposed above.

As mentioned in the theorical model, the depth of the phase reverse feature is affected by the excitation frequency. The phase of the induced current field at different depths was obtained with different excitation frequencies from 200 to 1000 Hz, as shown in Fig. 5a. The original phase of the induced current field is around the − 135° with different excitation frequencies. The phase goes down sharply as the excitation frequency increases. All the phases reverse at a specific depth with different excitation frequencies. The depth of the PRP drops with the increasing of the excitation frequency, as shown in Fig. 5b. For a lower excitation frequency, the PRP is in a deeper depth in the aluminum specimen. For example, when the excitation frequency is 300 Hz, the depth of the PRP is 4.89 mm. It means that the direction of the induced current field is in the reverse direction when the penetration depth is less than 4.89 mm. However, the direction of the induced current field is in the forward direction when the penetration depth is greater than 4.89 mm.

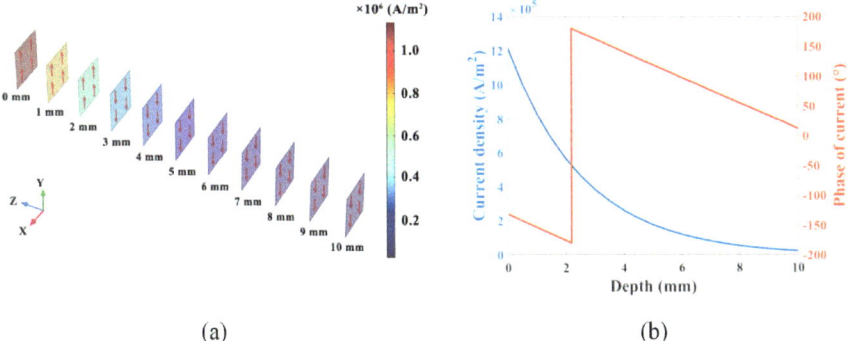

(a)   (b)

Fig. 4 Induced current field at different depths. **a** Directions of induced currents. **b** Induced current density and phase

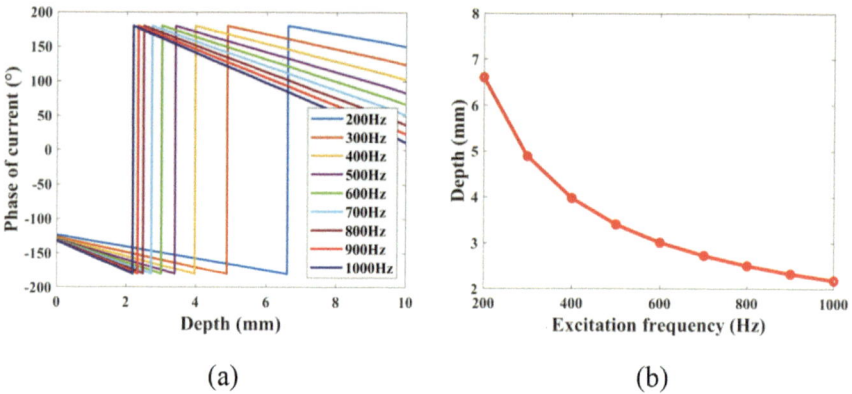

Fig. 5 Phase of induced current with different excitation frequencies. **a** Phase of induced current. **b** Depths of PRP

C. Characteristic Signal Analysis

To analyze the magnetic field response signals of cracks, a surface crack and a subsurface crack were set up in the simulation model. The size of the surface crack was 30 mm (Length) × 0.5 mm (Width) × 2 mm (Depth). The buried depth of the subsurface crack was 3 mm, and the length and width were the same as the surface crack. The 200 Hz and 1 kHz were selected as the frequencies of the excitation signal. According to the simulated results, the depths of the PRP were 2.19 mm (1 kHz excitation) and 6.6 mm (200 Hz excitation), respectively.

For the surface crack, the reverse phase current field deflects in the clockwise direction at one end of the surface crack with the excitation frequency of 200 Hz, as shown in Fig. 6a. Meanwhile, the reverse phase current field deflects in the anti-clockwise direction at another end of the surface crack. According to the Ampere's Law, the Bz shows a trough at one tip of the surface crack and a peak at another tip of the surface crack, as shown in Fig. 6c. Because the PRP of the 200 Hz is 6.6 mm, the surface crack is not affected by the phase reverse feature. When the excitation frequency is 1000 Hz (PRP is 2.19 mm), although the surface crack is located in the reverse phase and forward phase current fields at the same time, the reverse phase current field maily deflects in the same way around the surface crack, as shown in Fig. 6b. Thus, the Bz also shows a trough at one tip of the surface crack and a trough at another tip of the surface crack, as shown in Fig. 6c. Due to the different current density, the peaks of the Bz with 1000 Hz excitation frequency are stronger than that of the 200 Hz excitation frequency.

For the subsurface crack (Buried depth 3 mm), the reverse and forward phase induced currents are disturbed at the same time when the excitation frequency is 200 Hz, as shown in Fig. 7a. When the depth is less than the PRP (6.6 mm), the deflection direction of the induced current is in the clockwise direction at one end of the subsurface crack (reverse phase induced current field). However, When the depth is more than the PRP (6.6 mm), the deflection direction of the induced current is in

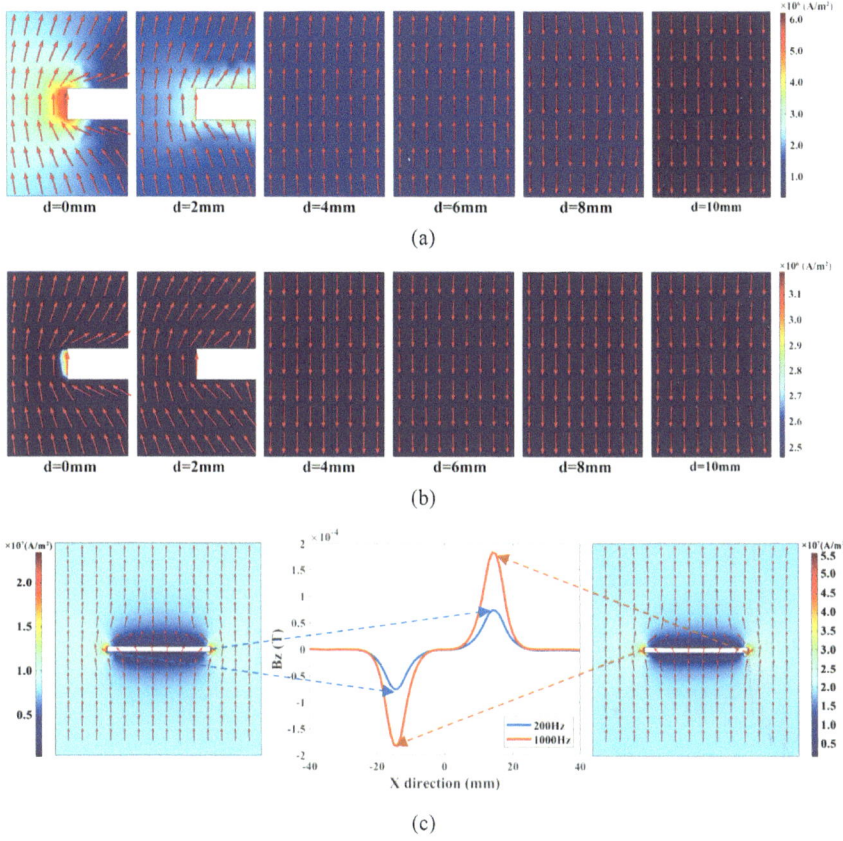

Fig. 6 Disturbed current field and distorted magnetic field of surface crack. **a** 200 Hz excitation frequency. **b** 1000 Hz excitation frequency. **c** Distorted magnetic field signal Bz

the anticlockwise direction at the same end of the subsurface crack (forward phase induced current field). Because the density of the reverse phase current is greater than that of the forward phase current, the Bz still shows a trough at one end and a peak at another end of the subsurface crack, as shown in Fig. 7c. However, when the excitation frequency is 1000 Hz, there is only the forward phase current field which is disturbed by the subsurface crack, as shown in Fig. 7b. This is because the buried depth of the subsurface crack is under the depth of the PRP (2.19 mm). Thus, the deflection direction of the induced current is in the anticlockwise direction at one end of the subsurface crack. As a result, the Bz shows a peak at one end and a trough at another end of the subsurface crack, which is opposite to the Bz signal excited by the 200 Hz sine signal, as shown in Fig. 7c.

When the excitation frequency is 500 Hz, the depth of the PRP is 3.4 mm. The edge of the subsurface crack is near the PRP area. The phase of the induced current changes from the reverse direction to the forward direction, which shows stray state,

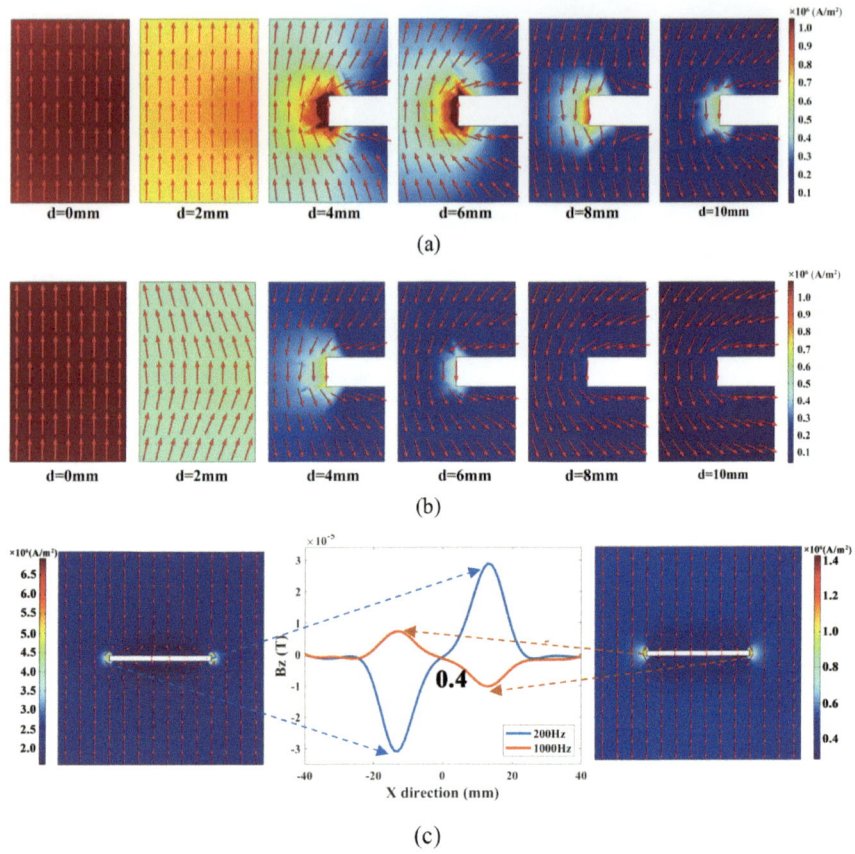

Fig. 7 Disturbed current field and distorted magnetic field of subsurface crack. **a** 200 Hz excitation frequency. **b** 1000 Hz excitation frequency. **c** Distorted magnetic field signal Bz

as shown in Fig. 8a. The stray currents cannot turn regularly. As a result, there is no obvious peak or trough in the Bz signal (Called transitional signal), as shown in Fig. 8b. In a conclusion, there are always peak and trough in the Bz signal for the surface crack with different frequency excitation signals. The peak and trough of the Bz signal can turn to the opposite direction for the subsurface crack with different frequency excitation signals. Especially, the transitional signal of the Bz is excited by a specific excitation frequency because of the stray current. Thus, the surface and subsurface cracks can be classified by the amplitude reversal phenomenon of the peak and trough of the Bz with multi-frequency excitation method. Because the transitional signal is caused by the stray current near the top side of the surface crack, the buried depth of the subsurface can be evaluated using the specific excitation frequency.

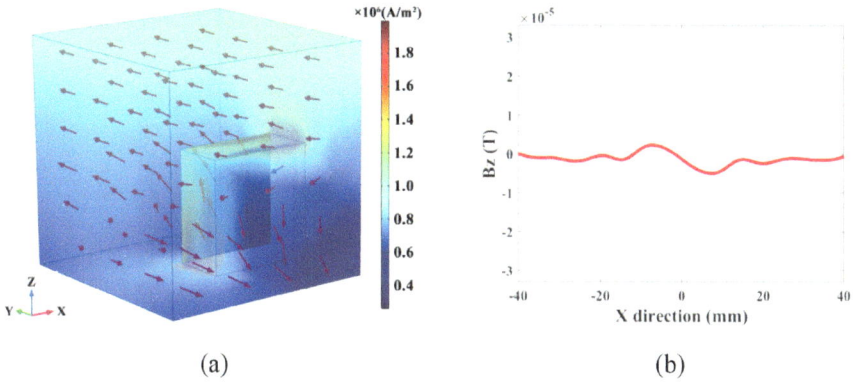

Fig. 8 Disturbed current field and distorted magnetic field of subsurface crack. **a** Stray current field. **b** Transitional signal

3 Multi-frequency ACFM Testing System

A. Multi-frequency Excitation Signal Synthesis

To verify the theoretical and simulated results, the multi-frequency excitation signal was synthesized. It has been proved that the 1 kHz was the optimal excitation frequency to detect surface cracks in the aluminum specimen [30]. Therefore, 1 kHz was set as the highest frequency component of the multi-frequency excitation signal. To get a penetration depth of 5 mm and above, 200 Hz was set as the minimum excitation frequency in this paper. The sinusoidal signals were added together with 200, 300, 400, 500, 600, 700, 800, 900, and 1000 Hz to generate the multi-frequency excitation signal. The amplitude of the sinusoidal signals was 1 V and the phase was 0°. The multi-frequency excitation signal was generated by LABVIEW software and output through an acquisition card (NI, USB6351) that was provided analog signal output function. The output multi-frequency excitation signal is shown in Fig. 9.

B. Probe and Testing system

As shown in Fig. 10a, a U-shape magnetic core, an excitation coil, a tunnel magneto resistance (TMR) sensor, and a signal processing circuit were packaged in the ACFM probe. The excitation coil (copper wire whose diameter was 0.15 mm) was wound around the U-shape core with 500 turns. The TMR sensor (Type: TMR2301, made by MULTI DIMENSION, China) was placed at the bottom of the probe, which was used to measure the Bz signal. The signal processing circuit was used to amplify the Bz signal and filter the interference noise.

As shown in Fig. 10b, the multi-frequency ACFM testing system included a probe, an acquisition card, a power amplifier, a DC power, and a personal computer (PC). The signal acquisition card was controlled by the PC to output the multi-frequency excitation signal. And then the excitation signal was amplified by the power amplifier. The amplified excitation signal was loaded on the excitation coil of the ACFM probe.

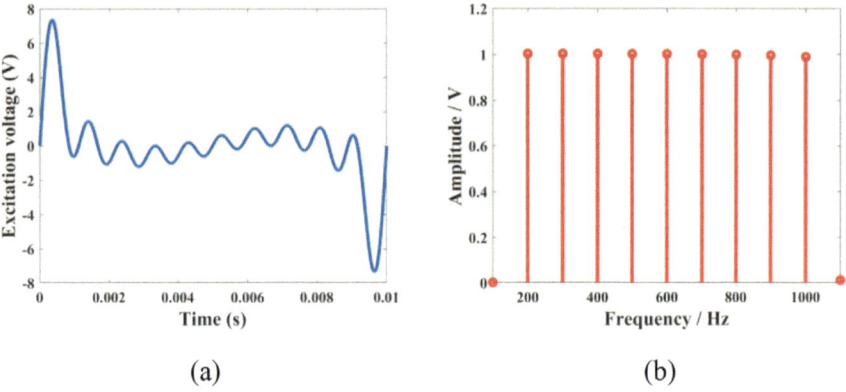

(a)   (b)

Fig. 9 Multi-frequency excitation signal. **a** Time domain signal. **b** Frequency domain signal

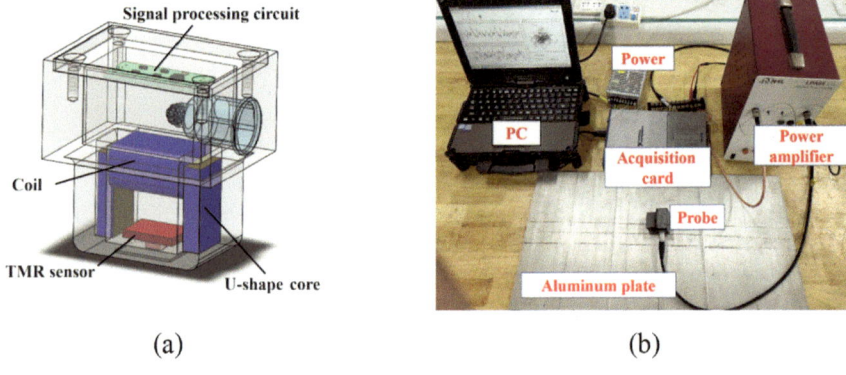

(a)   (b)

Fig. 10 Probe and testing system. **a** Detection probe. **b** Testing system

The uniform current field was excited into the aluminum plate by the probe. When a crack was presented, the induced current field would be disturbed. The disturbed current field made the space vertical magnetic field distorted (Bz) the above the crack. The Bz signal was picked up by the TMR sensor. Then the Bz signal was sent to the acquisition card. In the end, the Bz signal was processed and displayed by software in the computer.

4 Experiment

A. Classification of cracks

The specimen was an aluminum plate with a thickness of 10 mm, as shown in Fig. 11. There were five cracks (No. 1–5) with the same length (30 mm) and width (0.5 mm).

The five cracks could be considered as surface cracks or subsurface cracks (turned to another side). The depths of the surface cracks were 5 mm (No. 5), 6 mm (No. 4), 7 mm (No. 3), 8 mm (No. 2), and 9 mm (No. 1) respectively. The buried depths of the subsurface cracks were 1 mm (No. 1), 2 mm (No. 2), 3 mm (No. 3), 4 mm (No. 4), and 5 mm (No. 5) respectively.

The probe was set on the aluminum plate to measure the response signal of the Bz signal (one end of a surface crack), as shown in Fig. 12. There were 9 frequency components in the Bz signal. The amplitude of each frequency component in the frequency domain was set as the Bz amplitude at this location.

Firstly, the surface cracks were tested using the multi-frequency ACFM testing system. The peak value of each frequency component was extracted to obtain the Bz signals of the five surface cracks, as shown in Fig. 13. The Bz signals show negative peaks and positive peaks at the two ends of the surface crack. For a specific depth crack (for example 5 mm depth), the peak value of Bz increases as the excitation frequency goes down, as shown in Fig. 13a. This is because the amplitudes of different frequency components are different, as shown in Fig. 12b. For different depth surface cracks, all the Bz signals show peaks and troughs in the same direction. It indicates that the Bz signals of the surface cracks do not have the amplitude reversal phenomenon. This

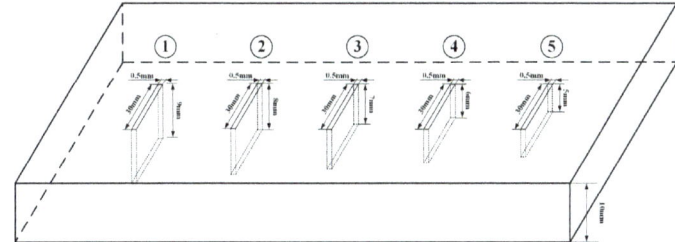

Fig. 11 Dimensions of the specimen and cracks

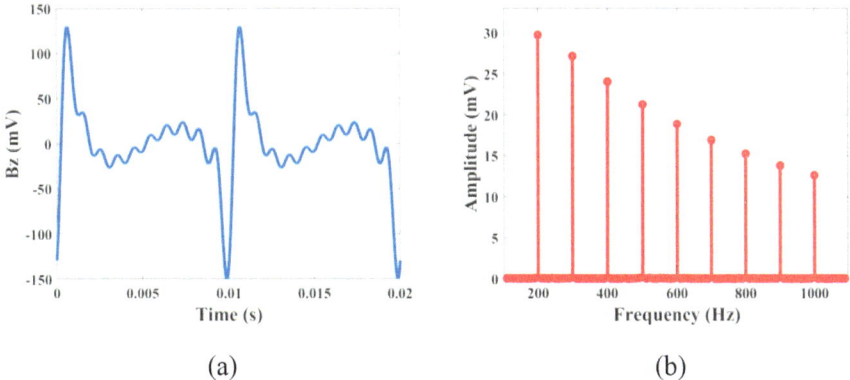

(a)   (b)

Fig. 12 Bz signal. **a** Bz time domain signal. **b** Bz frequency signal

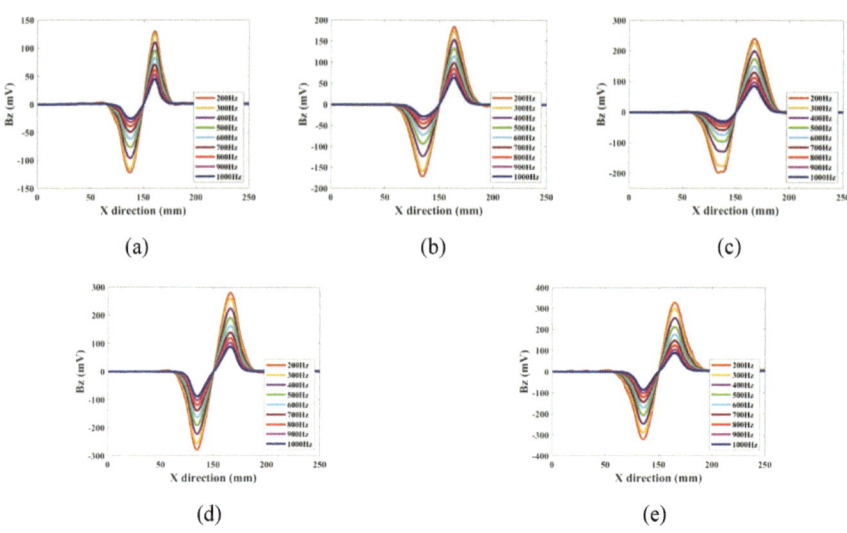

Fig. 13 Bz signals of the surface cracks. **a** 5 mm deep crack. **b** 6 mm deep crack. **c** 7 mm deep crack. **d** 8 mm deep crack. **e** 9 mm deep crack

is mainly because the Bz signals are influenced by the deflection direction of the reversal current. Although the forward phase current field is also disturbed around the bottom of the surface cracks, it only affects the magnitude of the Bz signal due to the weak current density.

Secondly, the subsurface cracks were tested by turning the aluminum plate to another side. The testing results of the subsurface cracks are shown in Fig. 14. It can be seen that the peak and trough of Bz signal changed to the opposite direction for each subsurface crack. It indicates that the Bz signals of the subsurface cracks show the amplitude reversal phenomenon. For the lower frequency component (for example 200 Hz in Fig. 14c), the Bz shows a trough and then a peak. For the higher frequency component (for example 1000 Hz in Fig. 14c), the Bz shows a peak and then a trough. This is because the 200 Hz excitation frequency has a deeper depth of the PRP (6.6 mm), which is larger than the buried depth (3 mm) of the subsurface crack. Although the reverse and forward phase induced currents are disturbed at the same time, the density of the reverse phase current is greater than that of the forward phase current. Thus, the peak and trough directions of the Bz signals keep the same all the time for the 1 mm to 5 mm buried depth subsurface crack when the excitation frequency is 200 Hz. The 1000 Hz excitation frequency has a shallower depth of the PRP (2.19 mm), which is less than the buried depth (3 mm). Only the forward current turns around the subsurface crack. The peak and trough of the Bz excited by 1000 Hz is opposite to that of the 200 Hz excitation signal.

All the peaks of the Bz goes down with the increases of the buried depth. This is because the attenuation of the current density in the depth direction. And it should be

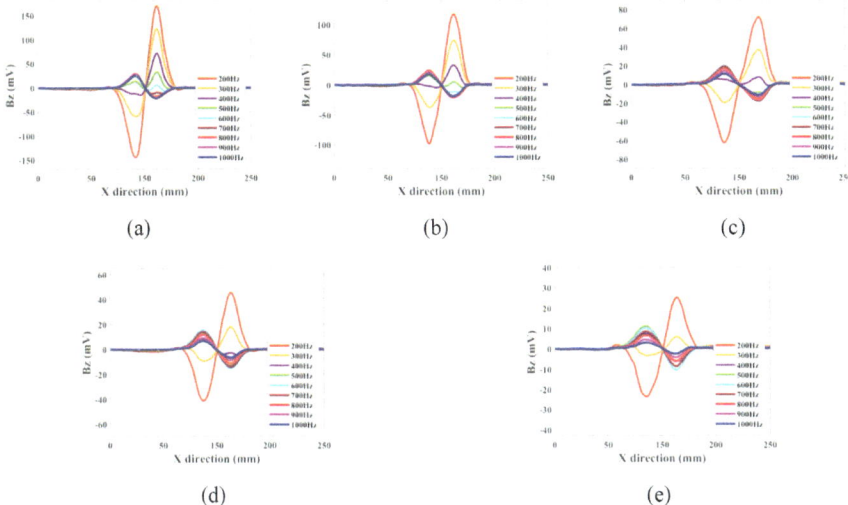

Fig. 14 Bz signals of the subsurface cracks. **a** 1 mm buried depth crack. **b** 2 mm buried depth crack. **c** 3 mm buried depth crack. **d** 4 mm buried depth crack. **e** 5 mm buried depth crack

noted that the transitional signals always exist in all buried depth cracks. The transitional signals are caused by the reversal frequency component near the buried depth of the subsurface crack. Thus, the reversal frequency component of the transitional signal can be used to evaluated the buried depth of the subsurface crack.

In conclusion, the Bz signal has one peak and one trough all the time whether it is a surface crack or a subsurface crack. However, the Bz signal shows different peaks and troughs for the subsurface crack with different excitation frequency components. With the increase of the excitation frequency, the amplitude reversal phenomenon will occur in the Bz signals. The results consistent with the previous theorical and simulated results. Thus, the surface and subsurface cracks can be classified by the amplitude reversal phenomenon of the Bz signal caused by the novel phase reversal feature using the multi-frequency ACFM technique.

B. Evaluation of subsurface cracks

Many scholars have proposed many methods to evaluate the depth of surface cracks. However, the buried depth of subsurface cracks is still a big challenge. Because the conventional amplitude characteristic is confused by the dimensionality and the buried depth of subsurface cracks, the buried depth cannot be evaluated well and truly. As mentioned above, there is transitional signal caused by the reversal frequency component near the buried depth of a subsurface crack. So the reversal frequency component can be used to evaluate the buried depth. Because the reversal frequency component is only related to the buried depth, it is not affected by the dimensionality of subsurface cracks. To find the reversal frequencies of the different buried depth subsurface cracks, the Bz signals of the five subsurface cracks were further processed using Eq. (8). All the frequency components were ploted in figures, as shown in

Fig. 15.

$$Bz_p = Bz/Bz_{\max} \tag{8}$$

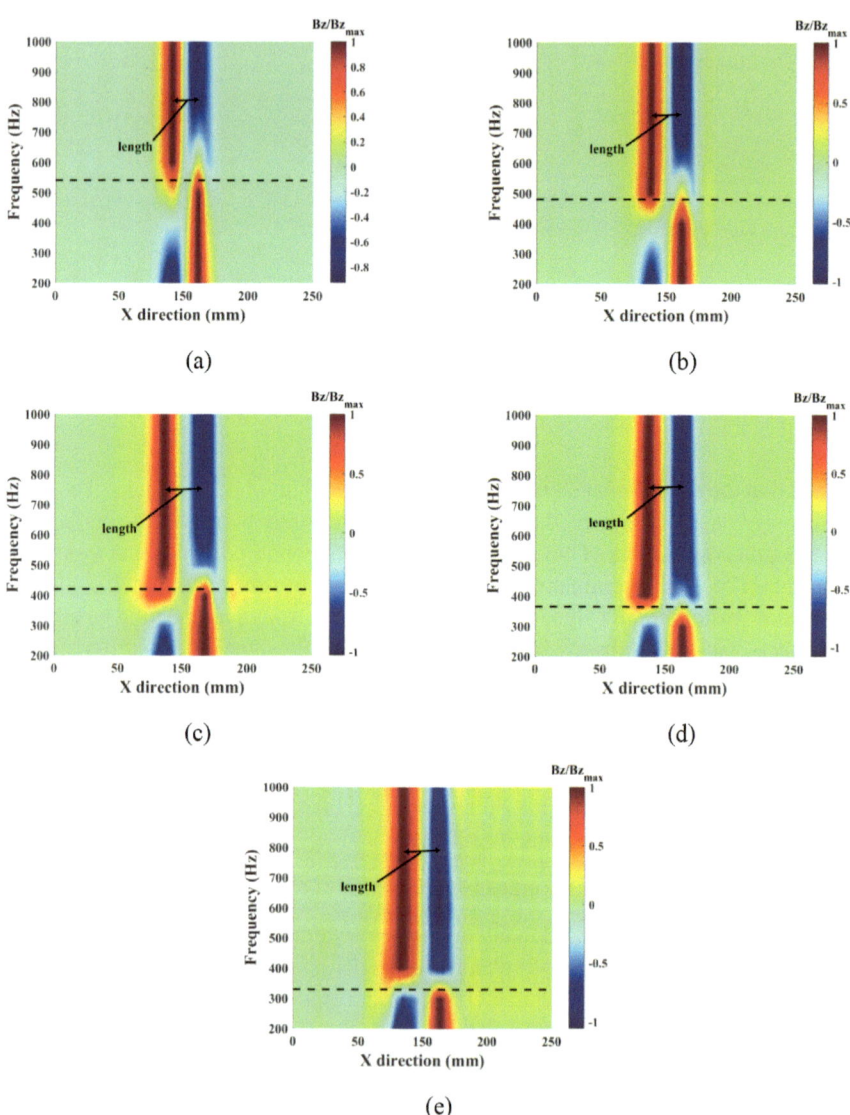

(a)

(b)

(c)

(d)

(e)

Fig. 15 Different frequency component signals of the subsurface cracks. **a** 1 mm buried depth crack. **b** 2 mm buried depth crack. **c** 3 mm buried depth crack. **d** 4 mm buried depth crack. **e** 5 mm buried depth crack

where Bz_p is the normalized Bz signal, Bz is unprocessed magnetic response field signal, Bz_{max} is the maximum value of the Bz signal.

As shown in Fig. 15, the distance between the peak and trough of the Bz signal indicates the crack length. The peak and trough of the Bz signals reverse at different reversal frequency components for the different buried depths cracks. The forward and reverse currents are mixed together to present the stray current state. Because the stray current caused by the reversal frequency component is a current area, it turns and gathers around the buried depth. The peak and trough of the Bz signals do not reverse at the same time as the frequency increases. Thus, the reversal frequency component is a transition zone, as shown in Fig. 15a. The two boundaries of the transition zone are the frequency of the peak reverse and the frequency of the trough reverse. To get a consistent and accurate reversal frequency component, the intermediate frequency component in the reversal frequency area was selected as the reversal frequency component to evaluate the buried depth. The reversal frequency components of the 5 subsurface cracks were marked with the dotted lines, as shown in Fig. 15.

The reversal frequency components were picked up, as shown in Fig. 16. The reversal frequency component decreases linearly with the increasing of the buried depth. It is consistent with the phase linear transmission of the induced current field, as shown in Fig. 4b. Thus, the buried depth can be evaluated by the reversal frequency component using the linear relationship function.

Three reversal frequency components of the subsurface cracks (1, 3 and 5 mm buried depth) were used to fit the linear relationship function. The relationship function between the buried depth and the reversal frequency component is shown in Eq. (9).

$$bd_e = -0.01775 \times f_r + 10.54 \tag{9}$$

$$E_{bd} = |bd_e - bd|/bd \tag{10}$$

where bd_e is the measured buried depth, f_r is the reversal frequency component. E_{bd} is the relative evaluation error, bd is the actual buried depth.

Fig. 16 Reversal frequency components of different buried depth cracks

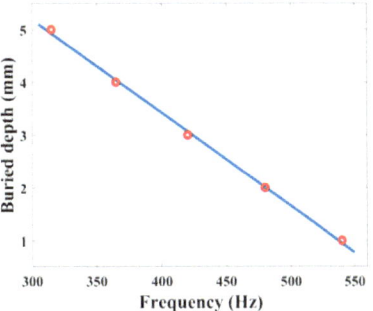

Table 1 Results of buried depth evaluation

	bd/mm	bd_e/mm	E_{bd} (%)
No. 2	2	2.02	1.0
No. 4	4	4.0612	1.53

The remaining two subsurface cracks (No. 2 and No. 4) can be evaluated using Eq. (9). The evaluated results are 2.02 mm and 4.0612 mm respectively. The relative evaluation error of the buried depth is defined in Eq. (10). The buried depth evaluated results and the relative evaluation errors are shown in Table 1. The relative evaluation errors of No. 2 and No. 4 cracks are 1.0 and 1.53% respectively. The buried depth of the subsurface crack can be evaluated accurately by the reversal frequency component.

In order to verify the generalisability of the proposed classification and evaluation method, a subsurface crack in another aluminium plate was detected using the developed system. The length of the subsurface crack was 14 mm, the width was 0.2 mm, and the buried depth was 4 mm. First of all, the probe was pushed to detect the subsurface crack. The detection result is shown in Fig. 17. The Bz signals of different frequencies show peaks and troughs. And as the frequency increases, the positions of the peak and trough exchange. It suggests that the Bz signals of the subsurface crack show the amplitude reversal phenomenon. It means that the proposed classification method is effective. Secondly, to evaluate the subsurface crack, the Bz signal was further processed using Eq. (8). The processed signal is shown in Fig. 17b and the reversal frequency of the subsurface crack is 353 Hz, which is marked in the figure. Finally, the subsurface crack was evaluated using Eq. (9), and the error was calculated using Eq. (10). The evaluated buried depth is 4.274 mm and the relative evaluation error is 6.85%. It shows that the novel phase reversal feature proposed in this paper is also effective and calculate for other specimens.

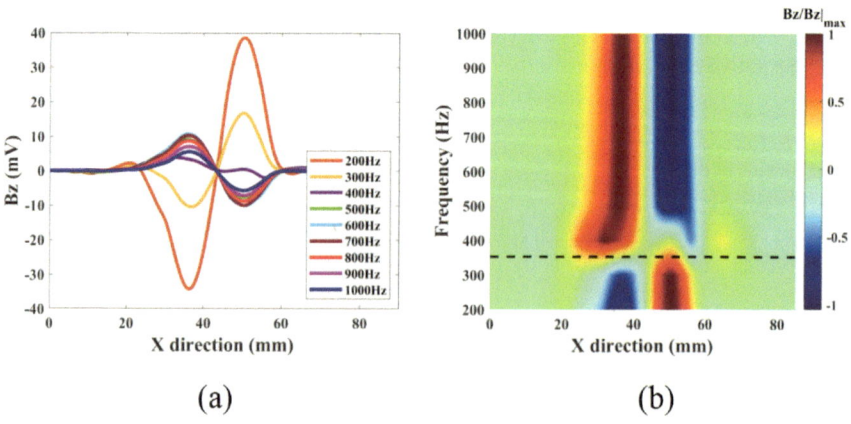

Fig. 17 Detection result of the subsurface crack. **a** Bz signals. **b** Bz_p signals

5 Conclusion

In this paper, the novel phase reversal feature is found to classify and evaluate cracks in the aluminum based on the multi-frequency ACFM technique. The physical principle of the phase reversal feature is developed. The distorted electromagnetic fields around the surface and subsurface cracks with different excitation frequencies are analyzed by the FEM model. The multi-frequency ACFM testing system is set up to test the surface and subsurface cracks. The results show that the peak and trough of the Bz signal caused by the subsurface crack can reverse with different excitation frequency component due to the phase reversal of the induced current field. The peak and trough of the Bz signal caused by the surface crack do not reverse. Thus, the surface and subsurface cracks can be classified by the amplitude reversal phenomenon of the Bz signal caused by the phase reversal feature. As the transitional signals of the Bz always exist in different buried depth caused by the stray current, the reversal frequency component is selected to evaluate the buried depth of the subsurface crack. The buried depth and the reversal frequency component have a good linear relationship. The buried depth of the subsurface crack can be evaluated accurately. Further work will focus on the evaluation of the buried depth with different lift-off distances and the evaluation of other complex subsurface defects.

References

1. D.R. Desjardins, T.W. Krause, A. Tetervak, L. Clapham, Concerning the derivation of exact solutions to inductive circuit problems for eddy current testing. NDTE Int. **68**, 128–135 (2014)
2. R.R. Hughes, High-sensitivity eddy-current testing technology for defect detection in aerospace superalloys. [Insertof Publication]: University of Warwick (2015)
3. X. Yuan, W. Li, G. Chen, X. Yin, W. Jiang, J. Zhao, J. Ge, Inspection of both inner and outer cracks in aluminum tubes using double frequency circumferential current field testing method. Mech. Syst. Sig. Pr. **127**, 16–34 (2019)
4. K. Baburaja, S.S. Teja, D.K. Sri, J. Kuldeep, V. Gowtham, Manufacturing and machining challenges of hybrid aluminium metal matix composites. IOP Conf. Ser. Mater. Sci. Eng. **225**(1), 12115 (2017)
5. Z. Li, H. Yang, X. Hu, J. Wei, Z. Han, Experimental study on the crush behavior and energy-absorption ability of circular magnesium thin-walled tubes and the comparison with aluminum tubes. Eng. Struct. **164**, 1–13 (2018)
6. M. Su, B. Young, L. Gardner, The continuous strength method for the design of aluminium alloy structural elements. Eng. Struct. **122**, 338–348 (2016)
7. C.P. Kohar, M. Mohammadi, R.K. Mishra, K. Inal, The effects of the yield surface curvature and anisotropy constants on the axial crush response of circular crush tubes. Thin Wall Struct. **106**, 28–50 (2016)
8. H.R. Hajibagheri, A. Heidari, R. Amini, An experimental investigation of the nature of longitudinal cracks in oil and gas transmission pipelines. J. Alloy. Compd. **741**, 1121–1129 (2018)
9. V.I. Pokhmurskii, I.M. Zin, V.A. Vynar, L.M. Bily, Contradictory effect of chromate inhibitor on corrosive wear of aluminium alloy. Corros. Sci. **53**(3), 904–908 (2011)

10. X.K. Yin, J.M. Fu, W. Li, G.M. Chen, D.A. Hutchins, A capacitive-inductive dual modality imaging system for non-destructive evaluation applications. Mech. Syst. Sig. Pr. **135**, 11458 (2020)
11. A. Bernieri, G. Betta, L. Ferrigno, M. Laracca, Crack depth estimation by using a multi-frequency ECT method. IEEE Trans. Instrum. Meas. **62**(3), 544–552 (2013)
12. J. Ge, N. Yusa, M. Fan, Frequency component mixing of pulsed or multi-frequency eddy current testing for nonferromagnetic plate thickness measurement using a multi-gene genetic programming algorithm. Ndt&E Int. **120**, 102423 (2021)
13. R. Mardaninejad, M.S. Safizadeh, Gas pipeline corrosion mapping through coating using pulsed eddy current technique. Russ. J. Nondestruct. **55**(11), 858–867 (2020)
14. V. Arjun, B. Sasi, B.P.C. Rao, C.K. Mukhopadhyay, T. Jayakumar, Optimisation of pulsed eddy current probe for detection of sub-surface defects in stainless steel plates. Sens. Actuat. A Phys. **226**, 69–75 (2015)
15. A.A.S.H. Sophian, G.Y. Tian, D. Taylor, J. Rudlin, A feature extraction technique based on principal component analysis for pulsed Eddy current NDT. NDTE Int. **1**, 37–41 (2003)
16. Y.H.Y. He, F.L.F. Luo, M.P.M. Pan, F.W.F. Weng, X.H.X. Hu, J.G.J. Gao, B.L.B. Liu, Pulsed eddy current technique for defect detection in aircraft riveted structures. NDTE Int **2**, 176–181 (2010)
17. S. Giguere, S.J.M. Dubois, Pulsed eddy current: finding corrosion independently of transducer lift-off. AIP Conf. Proceed. **509A**, 449–456 (2000)
18. G.Y. Tian, Y. Li, C. Mandache, Study of lift-off invariance for pulsed eddy-current signals. IEEE Trans. Magn. **45**(1), 184–191 (2009)
19. M. Fan, B. Cao, G. Tian, B. Ye, W. Li, Thickness measurement using liftoff point of intersection in pulsed eddy current responses for elimination of liftoff effect. Sens. Actuat. A Phys. **251**, 66–74 (2016)
20. A. Bernieri, G. Betta, L. Ferrigno, M. Laracca, S. Mastrostefano, Multifrequency excitation and support vector machine regressor for ECT defect characterization. IEEE Trans. Instrum. Meas. **63**(5), 1272–1280 (2014)
21. D.Z.D. Zhang, Y.Y.Y. Yu, C.L.C. Lai, G.T.G. Tian, Thickness measurement of multi-layer conductive coatings using multifrequency eddy current techniques(Article). Nondestruct. Test Eva **3**, 191–208 (2016)
22. S.A. Xie, L.A. Zhang, Y.A. Zhao, X.A. Wang, Y.A. Kong, Q.A. Ma, Z.A.C.M. Chen, T.A. Uchimoto, T.A. Takagi, Features extraction and discussion in a novel frequency-band-selecting pulsed eddy current testing method for the detection of a certain depth range of defects. NDTE Int. **12**, 102211 (2020)
23. J. Gao, M. Pan, F. Luo, D. Chen, Research on signal processing method of multi-frequency eddy current testing based on spectrum analysis. J. Electr. Measur. Instrum. **25**(1), 16–22 (2011)
24. J. Zhao, W. Li, X.A. Yuan, X. Yin, G. Chen, X. Li, J. Ding, X. Liu, Detection system development of drill pipe thread based on ACFM technique. IEEE Sens. J. **215**, 1 (2021)
25. X. Yuan, W. Li, G. Chen, X. Yin, J. Ge, W. Yang, J. Liu, W. Ma, Inner circumferential current field testing system with TMR sensor arrays for inner-wall cracks inspection in aluminum tubes. Measurement **122**, 232–239 (2018)
26. J. Ge, W. Li, G. Chen, X. Yin, J. Liu, Q. Kong, X. Yuan, Investigation of optimal time-domain feature for non-surface defect detection through a pulsed alternating current field measurement technique. Measur. Sci. Technol. **29**(1), 15601 (2017)
27. J.H. Ge, B.W. Hu, C.K. Yang, Investigation of the approximate decomposition of alternating current field measurement signals from crack colonies. Mech. Syst. Sig. Pr. **160**, 361 (2021)
28. X.A. Yuan, W. Li, G. Chen, X. Yin, J. Ge, Circumferential current field testing system with TMR sensor array for non-contact detection and estimation of cracks on power plant piping. Sens. Actuat. A Phys. **263**, 542–553 (2017)
29. J. Zhao, W. Li, J. Ge, X. Yuan, X. Yin, Y. Zhu, Z. Wang, Y. Liu, J. Li, Coiled tubing wall thickness evaluation system using pulsed alternating current field measurement technique. IEEE Sens. J. **20**(18), 10495–10501 (2020)

30. M.P.A. Papaelias, M.C.B. Lugg, C.A. Roberts, C.L.A. Davis, High-speed inspection of rails using ACFM techniques. NDTE Int. **4**, 328–335 (2009)
31. M. Smith, C. Laenen, Inspection of nuclear storage tanks using remotely deployed ACFMT. Insight **1**, 17–20 (2007)
32. J. Munoz, F. Marquez, M. Papaelias, Railroad inspection based on ACFM employing a non-uniform B-spline approach. Mech. Syst. Sig. Pr. **40**(2), 605–617 (2013)

Visual Reconstruction of Irregular Crack in Austenitic Stainless Steel Based on ACFM Technique

Abstract For the advantages of corrosion resistance, high toughness and plasticity, the austenitic stainless steel is widely used in the petrochemical special equipment and offshore structure. The austenitic stainless steel usually services in high temperature, high pressure and corrosive medium environment. Various types of irregular cracks are easily introduced in the surface of the austenitic stainless steel, which threats the safety of the structure. Due to the property of non-magnetic, weak conductivity and coarse grain, it is still a challenge to detect and evaluate irregular cracks in the austenitic stainless steel using the nondestructive testing (NDT) method. The visual reconstruction method is presented to detect and evaluate the irregular crack in the austenitic stainless steel based on the alternating current field measurement (ACFM) technique. The austenitic stainless steel irregular crack ACFM finite element simulation model is set up. The distorted electromagnetic field around the irregular crack is analyzed. The vertical direction magnetic field (Magnetic field perpendicular to specimen, called Bz) image gradient field visual reconstruction method is presented to reconstruct the surface profile of the irregular crack. The irregular crack testing experiments are carried out to verify the efficiency of the visual reconstruction method. The results show that the current field induced by the ACFM probe gathers at the tips of the irregular crack in the austenitic stainless steel. The gathered current makes the vertical direction magnetic field Bz distorted. The Bz image gradient field can reflect the location of the gathered current. The surface profile of the irregular crack can be imaged visually and evaluated accurately by the Bz image gradient field visual reconstruction method.

Keyword Visual reconstruction · Irregular crack · Austenitic stainless steel

X. Yuan et al., *Recent Development of Alternating Current Field Measurement Combine with New Technology*, https://doi.org/10.1007/978-981-97-4224-0_6

1 Introduction

Austenitic stainless steel possesses excellent corrosion resistance, high toughness, and machinability, making it widely used in various fields such as petrochemical equipment, marine equipment, and power generation equipment [1]. Generally, austenitic stainless steel serves in high-temperature, high-pressure, and corrosive environments, where various types of stress corrosion cracking (SCC), fatigue cracking, and intergranular corrosion are prone to occur on the surface of the structure, ultimately forming irregular crack defects [2, 3]. Irregular cracks can rapidly accumulate and propagate under external forces, leading to structural failure and posing a serious threat to structural safety and serviceability [4]. Therefore, conducting research on the visualization and reconstruction techniques of irregular cracks on the surface of austenitic stainless steel is of significant importance and engineering application value for timely grasping crack propagation morphology information and facilitating structural safety assessment and maintenance decision-making.

Austenitic stainless steel exhibits non-magnetic properties, weak electrical conductivity, and coarse grain size, which make conventional magnetic particle testing (MT) and magnetic flux leakage testing (MFL) techniques impractical. Ultrasonic testing (UT) is mainly used for internal defect detection and is not sensitive to surface cracking [5, 6]. Additionally, due to the coarse grain size and complex acoustic reflection signals within austenitic stainless steel, UT is not suitable for detecting surface irregular cracks [7]. Penetrant testing (PT) can be used for detecting small surface cracks and revealing the opening direction of surface cracks. However, PT requires thorough cleaning of the surface of austenitic stainless steel from adhering substances, oil contamination, and coatings, resulting in low operational efficiency and difficulties in on-site testing. Moreover, the penetrant agents can cause environmental pollution. Alternating current potential drop (ACPD) method can detect and monitor local structural crack detection by injecting current and using contact probes, but the probes need to penetrate the coating and directly contact the structure surface [8]. Eddy current testing (ET) is a widely used non-destructive testing technique for surface defect detection in structures. It relies on impedance analysis to detect and evaluate defects. However, ET is susceptible to lift-off disturbances and requires high surface smoothness of the structure [9, 10].

Alternating Current Field Measurement (ACFM) is an emerging electromagnetic non-destructive testing technique in recent years. It integrates the advantages of eddy current testing and alternating current potential drop method. By using an excitation coil to induce uniform current on the surface of a structure, ACFM can detect and evaluate surface cracks by measuring the distorted magnetic field above the defect when cracks occur, causing the current to accumulate at the crack endpoints and perturb the spatial magnetic field [11, 12]. The advantage of a uniform electric field makes the probe less susceptible to lift-off height and enables crack detection and evaluation without the need to remove attachments and coatings (up to a thickness of 10 mm). The ACFM mathematical model is precise and allows for accurate assessment of crack length and depth [13]. In addressing the issue of irregular crack detection and evaluation, Nicholson et al. applied ACFM technology to the detection of irregular rolling contact fatigue cracks on steel rails and proposed an evaluation method for inclined surface cracks [14]. Noroozi et al. utilized ACFM technology and fuzzy learning methods for arbitrary profile crack detection and reconstruction [15]. Ravan et al. developed a computational model for calculating the electromagnetic field perturbations around arbitrary profile cracks and solved for the analytical solution of the electromagnetic field inside the crack region using the finite difference method [16]. Pasadas et al. investigated the current perturbation patterns around irregular surface cracks on aluminum plate specimens under uniform current excitation and proposed a Tikhonov normalization surface crack visualization inversion method [17]. Li Yong et al. [18] achieved visualization and imaging display of buried depth defects in aluminum plates using pulse uniform eddy current technology and characteristic signal gradient fields. In previous research, our research group proposed a method for detecting arbitrarily oriented cracks using rotating alternating current electromagnetic fields, achieving high sensitivity detection of arbitrarily oriented cracks [19, 20]. While the aforementioned studies have laid the foundation for irregular crack detection and evaluation, there is limited research and reporting on the visualization and reconstruction methods of surface contour irregular cracks specifically in austenitic stainless steel.

In response to the issue of detecting and evaluating irregular cracks on the surface of austenitic stainless steel, this study proposes a method for visual reconstruction of such cracks based on the Alternating Current Field Measurement (ACFM) technique. By analyzing the electromagnetic field distortion patterns around irregular cracks through simulation models, the study introduces a technique for visualizing the surface contour of irregular cracks using the gradient field of the vertical magnetic field (Bz) image. Experimental tests are conducted to detect irregular cracks on austenitic stainless steel using ACFM, and the proposed visualization reconstruction method based on the Bz image gradient field is employed for precise imaging display and accurate assessment of the surface contour of irregular cracks.

2 An Irregular Crack Simulation Model

2.1 Simulation Model

An ACFM simulation model of an irregular crack on austenitic stainless steel is established using the ANSYS finite element simulation software, as illustrated in Fig. 1. The simulation model primarily consists of an excitation coil, a U-shaped magnetic core, a test specimen, and the irregular crack. The excitation coil is wound around the crossbeam of the U-shaped magnetic core with 500 turns, and the irregular crack is located beneath the excitation coil.

The simulation model and experimental specimen utilize austenitic stainless steel 316L for the test. The U-shaped magnetic core is made of manganese-zinc ferrite,

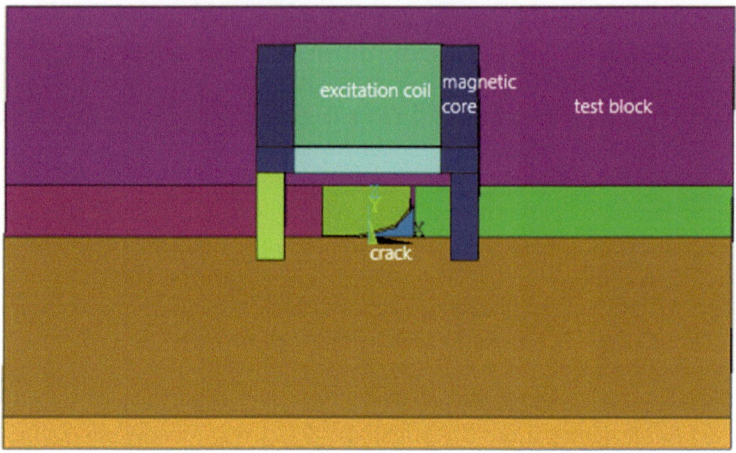

Fig. 1 ACFM simulation model of austenitic stainless steel irregular crack

Table 1 Parameters for the simulation model

Model parameters	Media environment	Exciting coil	U-shaped magnetic core	Stainless steel test block
Resistivity/(Ω·m)	0.5×10^4	1.7×10^{-8}	1.9×10^{-6}	7.4×10^{-7}
Relative permeability μ_r	1.0	1.0	2000	1.1

while the excitation coil is composed of copper wire. The remaining medium is air. The material parameters for the simulation model are provided in Table 1. The irregular crack consists of four segments, each with the same length of 20 mm and a depth of 3 mm. The crack angles are oriented at 0°, 30°, 60°, and 90° relative to the scanning direction.

2.2 Simulation Analysis of Electromagnetic Field

A sinusoidal excitation signal with a frequency of 2 kHz and an amplitude of 5 Vpp is applied to the excitation coil. The excitation coil induces a uniform current field on the surface of the austenitic stainless steel. The surface induction current vector map of the test specimen is extracted, as shown in Fig. 2. In the defect-free region, the induction current appears uniform. However, due to the presence of the irregular crack, the induction current accumulates at the endpoints and on both sides of the irregular crack.

In the defect-free region, the induction current is uniformly distributed, resulting in a vertical magnetic field Bz of 0. However, in the presence of an irregular crack,

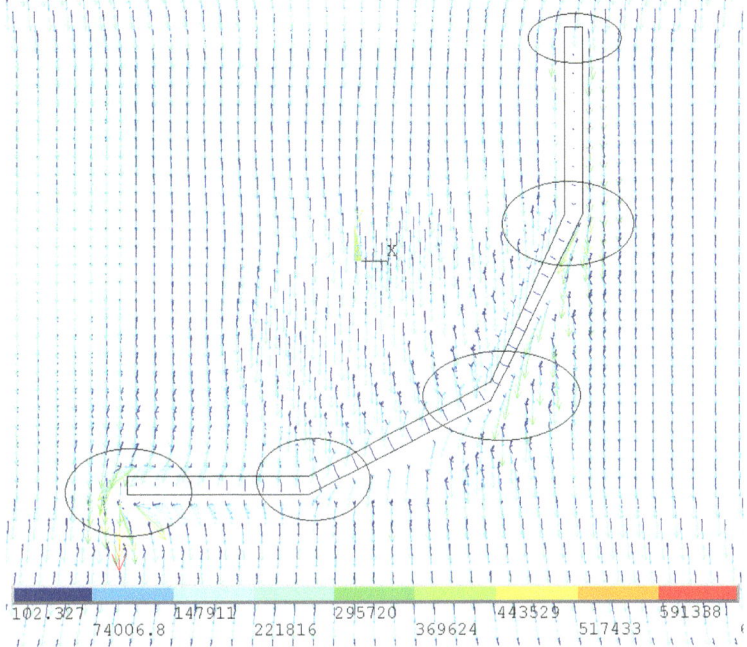

Fig. 2 Law of perturbation around irregular cracks by induction current

the accumulation of current causes spatial magnetic field distortion. The currents with different rotational directions lead to peak or trough values of Bz at the crack endpoints. The Bz image is extracted at a position 2 mm from the surface of the test specimen, as shown in Fig. 3. Bz exhibits peaks and troughs at different positions along the crack, and the distorted peak and trough positions align with the locations where the current accumulates at the endpoints of the irregular crack.

The simulation results indicate that the ACFM induction current can accumulate at the endpoints and on both sides of the irregular crack. This accumulation of current causes distortion in the vertical magnetic field Bz. The Bz image exhibits positive and negative peaks at the locations where the current accumulates, reflecting the surface topography information of the irregular crack.

Fig. 3 Vertical magnetic
field Bz image

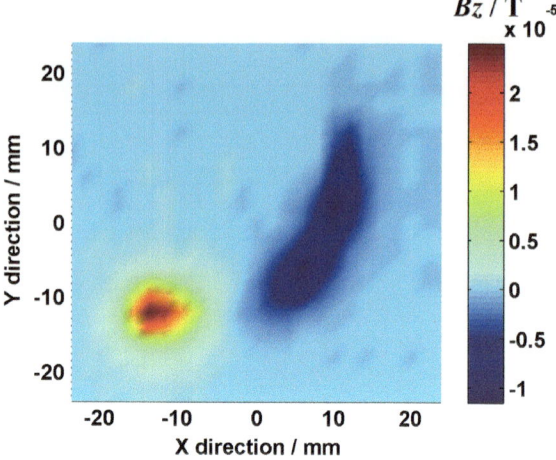

3   Visualization Reconstruction Method

3.1   *Gradient Field Algorithm*

The gradient field reflects both the magnitude and direction of a scalar field. It can be obtained by taking the gradient of the scalar field. The gradient field is defined as follows:

$$\boldsymbol{grad}\ u(x, y, z) = \left(\frac{\partial u}{\partial x}, \frac{\partial u}{\partial y}, \frac{\partial u}{\partial z}\right) = \nabla u(x, y, z) \tag{1}$$

In the formula, $u(x, y, z)$ is the quantity field, $\boldsymbol{grad}\ u(x, y, z)$ or $\nabla u(x, y, z)$ is referred to as the gradient field of the quantity field $u(x, y, z)$.

The positions of positive and negative peaks in the vertical magnetic field Bz reflect the surface topography information of the irregular crack. By calculating the gradient field of the Bz image, we can obtain information about the distorted peak positions and their orientations. As shown in Fig. 4, the gradient fields in the X-direction (probe scanning direction) and Y-direction (direction of the induction current, perpendicular to the scanning direction) are calculated from the Bz image. These gradient field images represent the feature signals in both directions (the gradient images are dimensionless and only represent relative changes in magnitude).

In Fig. 4a, the gradient field image in the X-direction of the Bz image represents the surface contour of the irregular crack. In Fig. 4b, the gradient field in the Y-direction of the Bz image reflects the contours on both sides of the irregular crack. The gradient field image in the X-direction of the Bz image is more suitable for reconstructing the surface contour of the irregular crack. Therefore, this paper proposes a visualization reconstruction method for the surface contour of irregular cracks using the gradient

Fig. 4 Bz image gradient
field signal

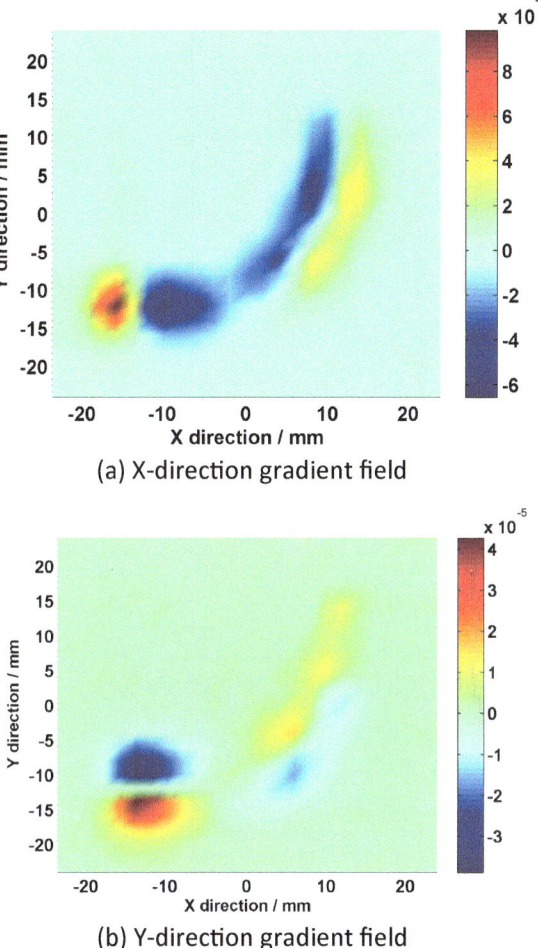

(a) X-direction gradient field

(b) Y-direction gradient field

field of the vertical magnetic field Bz image. The specific steps of the method are illustrated in Fig. 5.

Step 1: Calculate the X-direction gradient field of the vertical magnetic field Bz. Define the probe scanning direction as the X-direction and calculate the X-direction gradient field (GXBz) from the feature signal Bz.

Step 2: Determine the extrema. Determine the sign of the gradient field GXBz (PGXBz). If it is positive, proceed to the next step. If it is negative, multiply it by (-1).

Step 3: Remove negative background values. Check if GXBz is greater than 0. Keep the positive values and discard the negative values.

Step 4: Normalize the data. Normalize GXBz to the range of 0–1, obtaining the normalized signal representing the surface of the crack. Plot it as a color map.

Fig. 5 Visual reconstruction
method of Bz image gradient
field

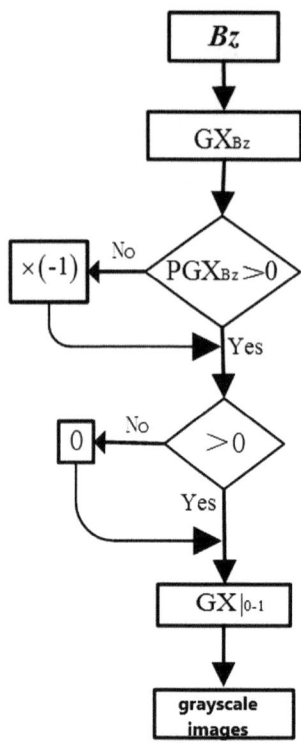

Step 5: Convert to a grayscale image. Convert the obtained normalized signal color map to a grayscale image, resulting in a visualization of the surface contour of the crack.

3.2   Simulation Results Visualization Refactoring

Using the algorithm described above, the second and third steps are applied to further process the X-direction gradient field of the vertical magnetic field Bz image in Fig. 4a. This results in a gradient field signal with the background field removed, as shown in Fig. 6a. The gradient field signal with the background field removed clearly depicts the visual contour of the crack surface. Next, in step four, the signal in Fig. 6a is normalized to the range of 0–1, resulting in a normalized signal color map shown in Fig. 6b. Finally, step five is applied to convert the normalized color map in Fig. 6b into a grayscale image, resulting in the visualization reconstruction of the irregular crack in the austenitic stainless steel, as shown in Fig. 6c. The visualization reconstruction result in Fig. 6c closely matches the contour of the irregular crack, indicating that the gradient field method using the vertical magnetic field Bz image can effectively visualize the surface contour of irregular cracks.

Fig. 6 Visual reconstruction simulation results of irregular crack in austenitic stainless steel l

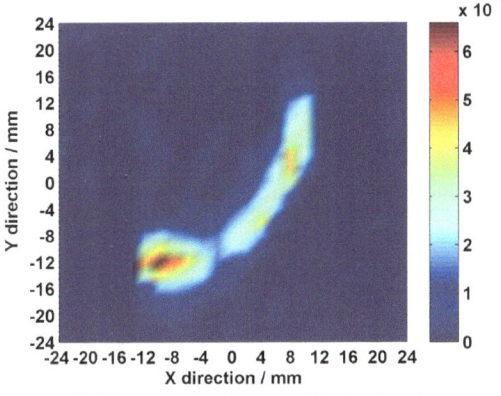

(a) Remove background field signals

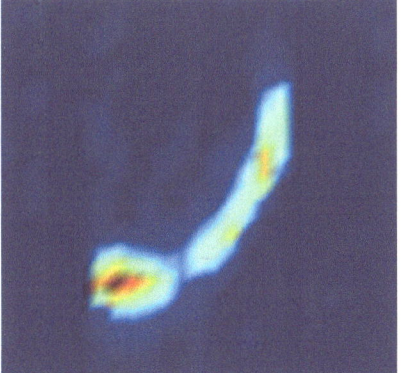

(b) Remove background field signals

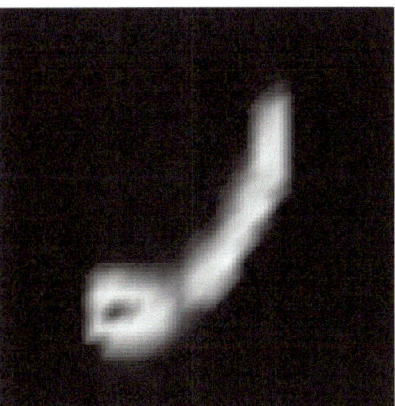

(c) Remove background field signals

4 Experimental Verification

4.1 Test System Construction

To verify the feasibility of the visualization and reconstruction method for the vertical magnetic field gradient (Bz) of irregular cracks in austenitic stainless steel, an ACFM experimental test system was constructed. The irregular cracks on the surface of the austenitic stainless steel specimen were visualized and reconstructed. As shown in Fig. 7, the ACFM test system mainly consists of a probe, a signal box, a control cabinet, and a three-axis platform. The probe includes an excitation coil, a U-shaped magnetic core, a magnetic field sensor, and an amplification and filtering circuit. The signal box generates a 2 kHz sinusoidal excitation signal with an amplitude of 5 Vpp, which is amplified and loaded onto the excitation coil inside the probe. The excitation coil induces a uniform current field on the surface of the specimen. When cracks are present, the current disturbance causes spatial magnetic field distortion. The magnetic field sensor (tunneling magnetoresistance magnetic field sensor) inside the probe picks up the distorted magnetic field signal. After initial processing by the amplification and filtering module inside the probe, the signal is further amplified by the processing module inside the enclosure. Finally, the magnetic field sensor detects the analog signal, which is then converted to a digital signal by the signal acquisition module and sent to the computer. The computer program performs digital filtering, phase-locked amplification, and averaging on the signal to obtain the characteristic signal Bz [21, 22]. The PLC inside the control cabinet controls the three-axis platform to move in a grid pattern along the surface of the specimen. The platform drives the probe to extract the vertical magnetic field Bz above the specimen using a step-by-step method, and finally, the Bz image of the scanning area is plotted.

4.2 Experimental Test

As shown in Fig. 8b, a 316L austenitic stainless steel specimen was used for the experiment. The specimen surface was processed using electrical discharge machining to create irregular cracks consisting of four sections. The crack length was 30 mm, and the crack angles with respect to the scanning direction were 0°, 30°, 60°, and 90°, as shown in Fig. 8. The crack width was 0.5 mm, and the crack depth was 3.0 mm.

The probe, driven by the three-axis platform, performed grid scanning of the irregular crack area. The probe was lifted to a height of 2 mm, and the scanning area was 100×100 mm^2 with a scanning step size of 0.5 mm. After completing the scanning, the Bz image was obtained, as shown in Fig. 9. The vertical magnetic field Bz exhibited positive and negative peak values in the area of the irregular cracks, with strong peak signals observed at the endpoints and both sides of the cracks.

The processing of Fig. 9 using the gradient field visualization and reconstruction method for the Bz image is performed in step one. This step involves obtaining the

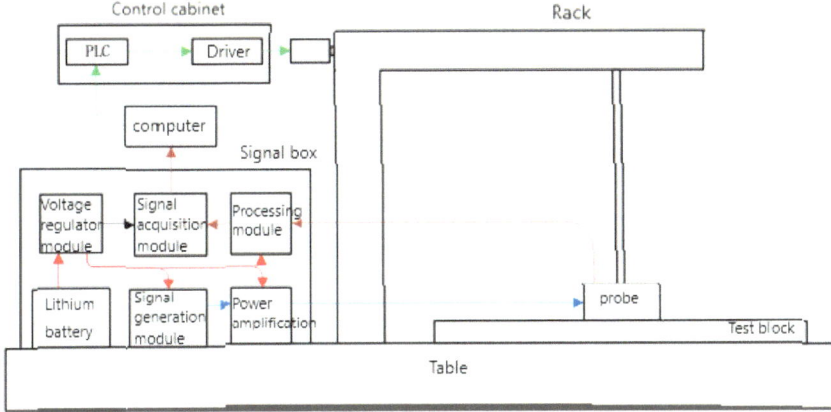

(a) ACFM test system block diagram

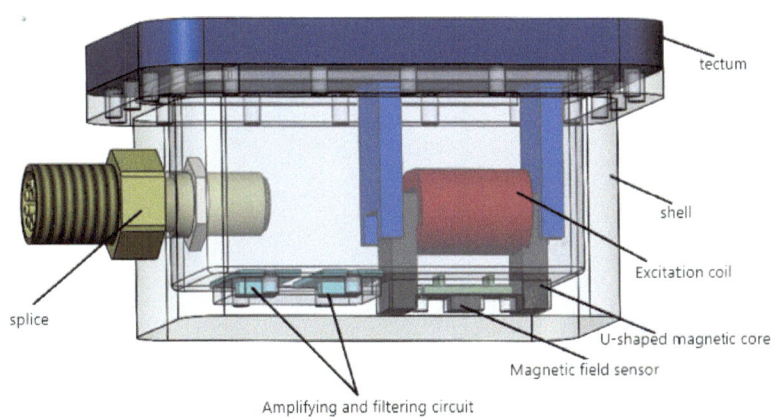

(b) Probe structure diagram

(c) Photo of ACFM test system

Fig. 7 ACFM test system

Fig. 8 Austenitic stainless
steel with irregular crack test
block

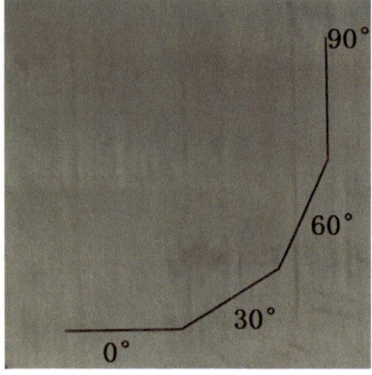

Fig. 9 Experimental results
of vertical magnetic field Bz

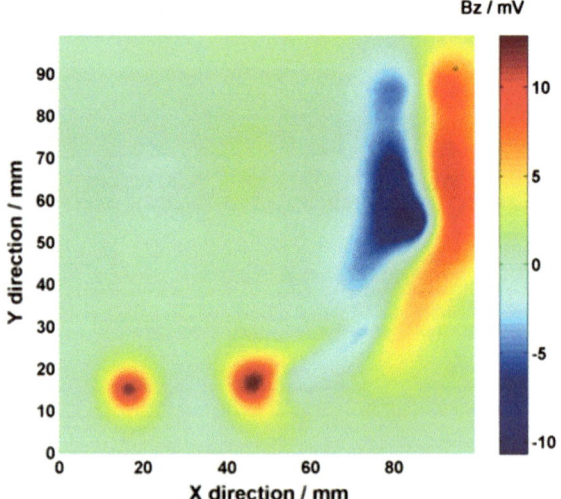

X-direction gradient field signal image of the Bz image. The positions of magnetic field distortion peaks are connected by lines, reflecting the contour direction of the irregular crack surface, as shown in Fig. 10a. In step two, step three, and step four of the visualization and reconstruction method, further processing is applied to Fig. 10a, resulting in a normalized gradient field image with the background removed, as shown in Fig. 10b. Step five is then applied to Fig. 10b, resulting in a visual reconstruction of the surface contour of the irregular crack, as shown in Fig. 10c. The visual reconstruction image reflects the positions of the crack endpoints and surface contours. Due to the current concentration effect at the crack endpoints, the vertical magnetic field Bz is stronger at the endpoint positions. The calculation results of the Bz image gradient field reflect the positions of extreme signals. Since the extreme signals are stronger and can overshadow the image of the intermediate crack region, they provide particularly prominent information about the positions of the crack endpoints.

Fig. 10 Experimental
results of visual
reconstruction of irregular
cracks in austenitic stainless
steel

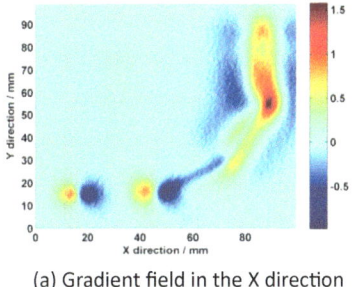

(a) Gradient field in the X direction

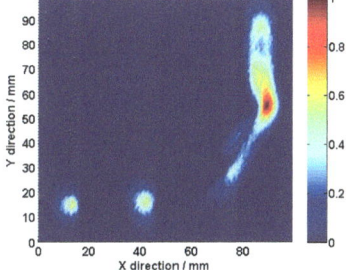

(b) Normalized image by removing background gradient field

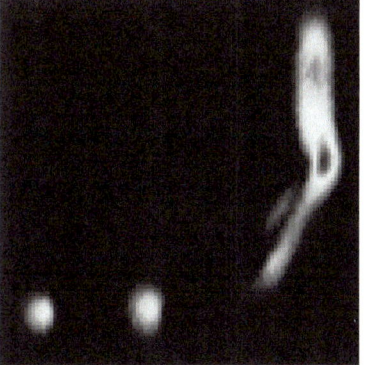

(c) Visual reconstruction results

4.3 *Reconstruction Accuracy Evaluation*

Based on the peak positions in the visualized image of the surface contour of the
irregular crack in Fig. 10c, the irregular crack can be divided into four sections. The
coordinates of the endpoints of each crack section are as follows: (13.5, 16.0), (42.0,
17.0), (75.0, 29.0), (89.0, 56.0), (86.5, 87.0). Using these endpoint coordinates, a
plot of the crack endpoints and their directions is created, resulting in an accurate
assessment of the irregular crack on the surface of the austenitic stainless steel, as
shown in Fig. 11.

Fig. 11 Evaluation results
of irregular cracks

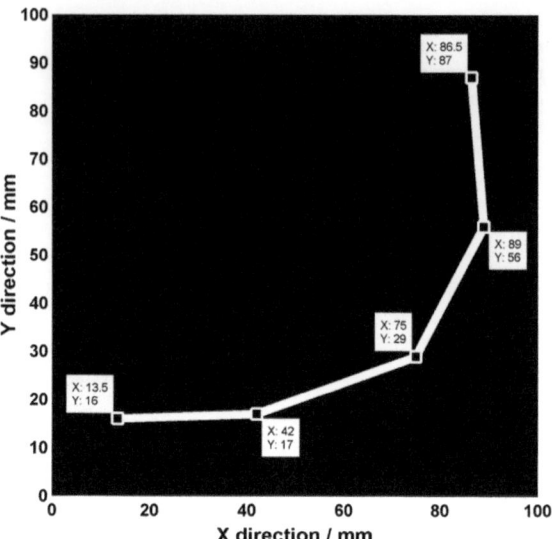

Based on the coordinates of the crack endpoint positions, the length and angle of each crack section can be calculated. The calculated lengths and angles for each section are as follows: Length = 28.5 mm, Angle = 2.0°. Length = 35.1 mm, Angle = 20.0°. Length = 30.4 mm, Angle = 62.6°. Length = 31.1 mm, Angle = 85.4°. By comparing these values with the true dimensions and angles of the irregular cracks on the surface of the austenitic stainless steel specimen, we can see that the maximum reconstruction error in the length of the irregular cracks is 5.1 mm, and the maximum error in the angle is 10°. This indicates a relatively high level of assessment accuracy in the visualization and reconstruction of the irregular crack lengths and angles.

5   Conclusions

(1)   This study simulated the electromagnetic field distortion around irregular cracks in austenitic stainless steel. The results showed that the ACFM-induced currents can accumulate at the endpoints and sides of irregular cracks, causing distortion in the vertical magnetic field Bz. The Bz exhibits peak and valley values at the positions of the crack endpoints, reflecting the information of the irregular crack endpoints and contours.

(2)   The gradient field visualization and reconstruction method using the vertical magnetic field Bz image was applied to reconstruct the ACFM simulation results of irregular cracks in austenitic stainless steel. The results showed that the X-direction gradient field image of the Bz image can present the surface contour of the irregular crack. By removing the background, normalizing the image, and

converting it to grayscale, a clear visualized image of the surface contour of the irregular crack can be obtained.

(3) An ACFM testing system was constructed to conduct experiments on the detection of irregular cracks in austenitic stainless steel, and the feasibility of the gradient field visualization and reconstruction method using the vertical magnetic field Bz image was validated. The results showed that the X-direction gradient field image of the Bz image can present a visualized image of the surface contour of the irregular crack, exhibiting extreme values at the crack endpoint positions. Through evaluation of the visualized results, the maximum error in crack length was 5.1 mm, and the maximum error in angle was 10°, validating the feasibility of the method. The research results of this study have strong guiding significance and engineering application value for the detection and evaluation of irregular cracks on the surface of austenitic stainless steel.

References

1. S. Shicheng, *Microstructure and mechanical properties of high nitrogen nickel-free austenitic stainless steel* (Jilin University, Changchun, 2014)
2. F. Ma, L. Pan, L. Zhang et al., Structure and property of AISI 316 and 304 austenitic stainless steel after low-temperature gas carburizing. Mater. Protect. **47**, 68–71 (2014)
3. Y. Zhao, J. Guan, F. Liu et al., Microstructure characterization of oxides formed on stainless steel with attachments of Fe–Zn intermetallics in high temperature water. Corr. Sci. Protect. Technol. **29**(6), 603–608 (2017)
4. P.C. Paris, F. Endorgan, A critical analysis of crack propagation laws. J. Basic Eng. **85**(4), 528–534 (1963)
5. G.S. Shelikhov, Y.A. Glazkov, On the improvement of examination questions during the nondestructive testing of magnetic powder. Russian J. Nondestruct. Test. **47**(2), 112–117 (2011)
6. J.B. Wu, H. Fang, X.M. Huang et al., An online MFL sensing method for steel pipe based on the magnetic guiding effect. Sensors **17**, 2911 (2017)
7. W. Li, Q. Liu, J. Zhang, Effect of microstructure of parent Metal and weld metal of austenitic stainless steel pipe on ultrasonic testing. Phys. Test. Chem. Anal. A Phys. Test. **51**(1), 22–26 (2015)
8. H. Saguy, D. Rittel, Alternating current flow in internally flawed conductors: a tomographic analysis. Appl. Phys. Lett. **89**(9), 094102 (2006)
9. G.Y. Tian, A. Sophian, Reduction of lift-off effects for pulsed eddy current NDT. NDTE Int. **38**, 319–324 (2005)
10. B. Gao, Y.Z. He, W.L. Woo et al., Multi-dimensional tensor-based inductive thermography with Multiple physical fields for offshore wind turbine gear inspection. IEEE Trans. Ind. Electr. **63**(10), 6305–6315 (2016)
11. X. Yuan, W. Li, G. Chen et al., Two-step interpolation algorithm for measurement of longitudinal cracks on pipe strings using circumferential current field testing system. IEEE Trans. Ind. Inform. **14**, 394–402 (2018)
12. W. Li, X. Yuan, G. Chen et al., Research on the detection of surface cracks on drilling riser using the chain alternating current field measurement probe array. J. Mech. Eng. **53**(8), 8–15 (2017)
13. A.M. Lewis, D.H. Michael, M.C. Lugg et al., Thin-skin electromagnetic fields around surface-breaking cracks in metals. J. Appl. Phys. **64**(8), 3777–3784 (1988)

14. G.L. Nicholson, C.L. Davis, Modeling of the response of an ACFM sensor to rail and rail wheel RCF cracks. NDTE Int. **46**(1), 107–114 (2012)
15. A. Noroozi, R.P.R. Hasanzadeh, M. Ravan, A fuzzy learning approach for identification of arbitrary crack profiles using ACFM technique. IEEE Trans. Magnet. **49**(9), 5016–5027 (2013)
16. M. Ravan, S.H.H. Sadeghi, R. Moini, Field distributions around arbitrary shape surface cracks in metals, induced by high frequency alternating current carrying wires of arbitrary shape. IEEE Trans. Magnet. **42**(9), 2208–2214 (2006)
17. D.J. Pasadas, A.L. Ribeiro, T. Rocha et al., 2D surface defect images applying Tikhonov regularized inversion and ECT. NDTE Int. **80**, 48–57 (2016)
18. Y. Li, B. Yan, L. Da et al., Gradient-field pulsed eddy current probes for imaging of hidden corrosion in conductive structures. Sens. Actuat. A Phys. **238**, 251–265 (2016)
19. W. Li, X.A. Yuan, G.M. Chen et al., High sensitivity rotating alternating current field measurement for arbitrary-angle underwater cracks. NDTE Int. **79**, 123–131 (2016)
20. W. Li, G. Chen, Defect visualization for alternating current field measurement based on the double u-shape inducer array. J. Mech. Eng. **45**(9), 233–237 (2009)
21. X. Yuan, W. Li, G. Chen et al., Circumferential current field testing system with TMR sensor array for non-contact detection and estimation of cracks on power plant piping. Sens. Actuat. A Phys. **263**, 542–553 (2017)
22. X. Yuan, W. Li, G. Chen et al., Inspection of both inner and outer cracks in aluminum tubes using double frequency circumferential current field testing method. Mech. Syst. Sig. Process. **127**, 16–34 (2019)

Visual ACFM System Modeling and Optimization for Accurate Measurement of Underwater Cracks

Abstract Alternating current field measurement (ACFM) has been proven to be a promising technique for sizing cracks on underwater structures. Precision and visualization are two big challenges for measuring underwater structure cracks, which is constrained by signal attenuation, interference factor and lags of signal processing. In this paper, an optimum visual ACFM (VACFM) system based on time-constant method (TCM) is proposed for visualization and accurate measurement of underwater structure cracks. Based on optimized results by finite element model (FEM), a novel probe and signal processing system are developed and sealed. Results obtained from crack inspection experiment with TCM show that the VACFM system is able to size the lengths and visually display the profiles of underwater structure cracks. The measuring errors of the crack length increase as the scanning speed increases; it is less than 10% at scanning speed of 3.81 mm/s, and smaller to longer cracks.

Keywords ACFM · Underwater · Visualization and accurate measurement · Crack

1 Introduction

In the last few decades, the number of subsea oil and gas key equipment, such as platforms and deepwater pipelines, has increased dramatically with the development of subsea oil and gas exploitation industries [1]. The development of underwater equipment makes it important to develop an intelligent and accurate cracks detection technology and system for discovering the defects and eliminating hidden dangers in underwater structures [2–4].

The conventional ACFM is used to detect and size the crack in offshore welded structures [5, 6]. The theoretical model of ACFM is shown in Fig. 1. The ACFM probe induces alternating current into the surface of the metal slab. If a surface-breaking crack is present, the current flows around ends of the crack. The changed current disturbs magnetic field above the crack. The magnetic field component, denoted as Bx (parallel to the slab), produces a deep trough in X direction, which contains depth information of the crack. Meanwhile, the magnetic field component, denoted as Bz

© The Author(s) 2024

X. Yuan et al., *Recent Development of Alternating Current Field Measurement Combine with New Technology*, https://doi.org/10.1007/978-981-97-4224-0_7

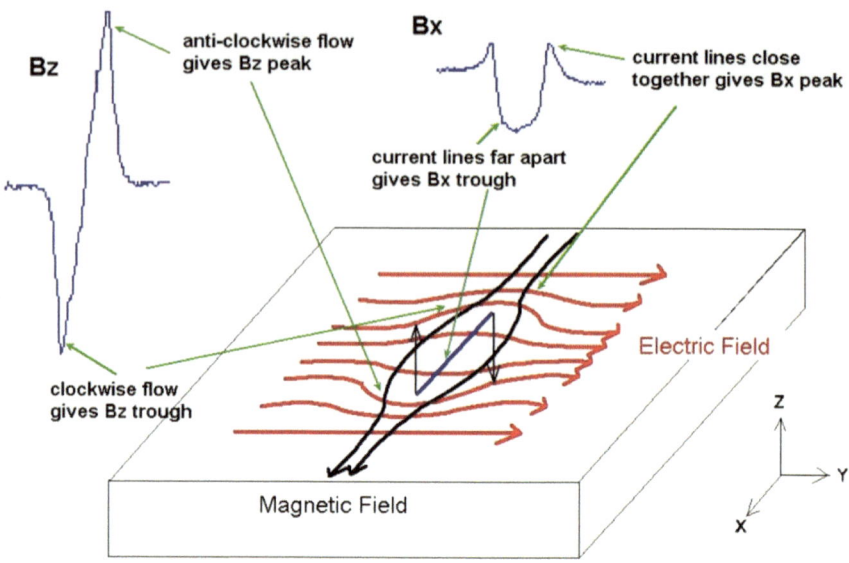

Fig. 1 The theoretical model of ACFM

(perpendicular to the slab), shows a peak and trough to the ends of the crack, which is indicative of the crack length [7, 8]. Because of the advantages of non-contact, fewer requirements of surface preparation, fast defects recognition and quantification, ACFM has become a promising alternative non-destructive test (NDT) technique for the conventional magnetic particle and penetrant testing methods [9].

Since the oil exploitation advances to abyssal sea, the demand for precise and high reliability underwater ACFM probe is increasing sharply [10]. For the existing challenges like signal attenuation and distortion in long-distance transmission [11], other interfering signals in the sea [12] and the lag of signal comparing with scanning speed for complex processing program [13], they are all critical factors to test the underwater crack precisely. Although ACFM does not need calibration, detection accuracy is still a big challenge in deepwater for interference signals [14]. In the last few years, commercially available ACFM technology has already offered visual results [15]. However, the conventional ACFM system cannot reconstruct the profile of crack visually and immediately, which brings difficulties to the operators to evaluate the test results [16].

In this work, a simulation model of underwater VACFM probe is built. The optimal excitation current, lift-off and structure are selected from simulation results. The underwater VACFM system is set up and tested by the crack inspection experiments in underwater environment. At the last part, a TCM is introduced to improve precision of the detected crack length.

2 Underwater VACFM

The schematic of underwater VACFM system is shown in Fig. 2. The system consists of two parts, underwater component and topside component. The underwater component of VACFM integrates excitation circuit, amplifier circuit, excitation coil, detecting sensor and conditioning circuit, which makes the system smaller and easier to handle. The topside component includes A/D acquisition card, DC power supply and personal computer.

In order to reduce the signal attenuation in long-distance transmission between topside and underwater, the excitation circuit, amplifier circuit, excitation coil, detecting sensor and conditioning circuit (amplification and filtering) are encapsulated the underwater probe. The excitation circuit produces driving signals. Signals are transferred to excitation coil through the power amplifier circuit. The excitation coil induces alternating current into the surface of sample under test (SUT). The detecting sensor pikes up the distorted magnetic field above the SUT surface. The signals are sent to topside after signal conditioning by the conditioning circuit. Then the signals are converted into digital signals by a A/D acquisition card and sent to PC at last. The intelligent identification software in PC will display the signals and identify defects.

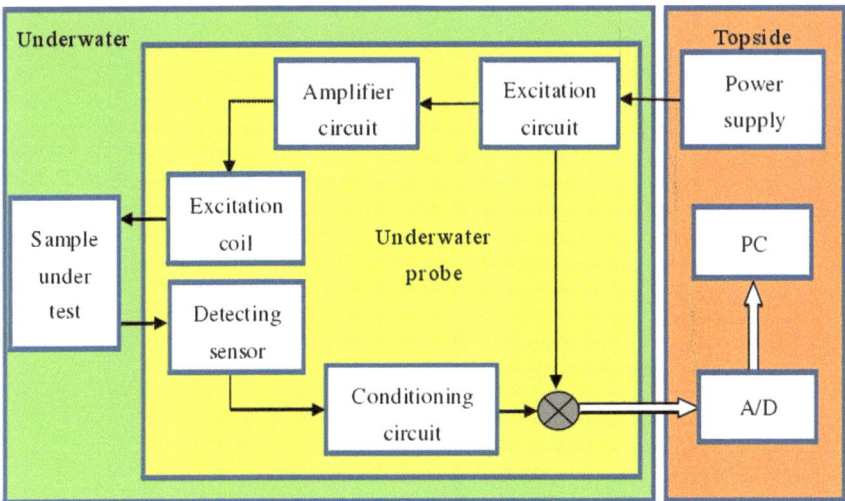

Fig. 2 The underwater VACFM system

3 Analysis and Optimization for Probe Parameters

3.1 Model Development in ANSYS

To optimize parameters of underwater VACFM probe, a FEM model is set up using ANSYS software, as shown in Fig. 3a. The model includes an alternating current-carrying coil with a U-shaped ferrite core, encapsulation shell and metal slab. The medium in the shell is air while the outside surrounding medium is seawater. A rectangular crack is introduced on slab surface in X direction. The dimensions of the model are shown in Table 1 and the characteristic parameters are shown in Table 2.

The current density is extracted on slab surface, as shown in Fig. 3b. Two eddy areas appear right under the legs of the U-shaped ferrite core; meanwhile a uniform current area [17] is generated between two legs on the slab. The uniform currents gather at ends of the crack and become sparse in the middle of the crack. A path is defined along X direction above the crack at the height of 2 mm from -0.03 to 0.03 m. The magnetic field is picked up on the path with 100 points, as shown in Fig. 3c. There is a peak (the absolute value is a trough) in Bx and a peak and trough in Bz. The peak of Bx lies in the middle of the crack, meanwhile the peak and trough of Bz locate at ends of the crack.

3.2 Excitation Currents

The induced current intensity on SUT surface is affected by the excitation current. When the excitation current is weaker, the current intensity becomes smaller, which is unfavorable for detecting the signal. Due to the power constraints, the excitation current should not be too big. Figure 4a and b shows the Bz and Bx with different excitation current. It can be seen, the excitation current increases as the amplitude of Bx and Bz increases. The maximum distortion (ΔMax) of Bx and Bz are showed in Fig. 4c respectively, which are both grow linearly with the excitation current. The sensitivity of Bx and Bz are given as follows [18].

$$\xi_x = \frac{\Delta Bx_{max}}{Bx_0} \tag{1}$$

$$\xi_z = \frac{\Delta Bz_{max}}{Bx_0} \tag{2}$$

ΔBx_{max} and ΔBz_{max} are the maximum distortion of Bx and Bz above defect. Bx_0 is the amplitude of Bx signal without defects. ξ can reduce detection errors and improve the signal-to-noise ratio, which affects the measurement accuracy of the VACFM system [19]. The ΔMax and sensitivity of Bx and Bz are showed in Table 3.

Fig. 3 The results of simulation. **a** The FEM of underwater VACFM probe. **b** The current density on the surface of the slab. **c** The magnetic field above the crack

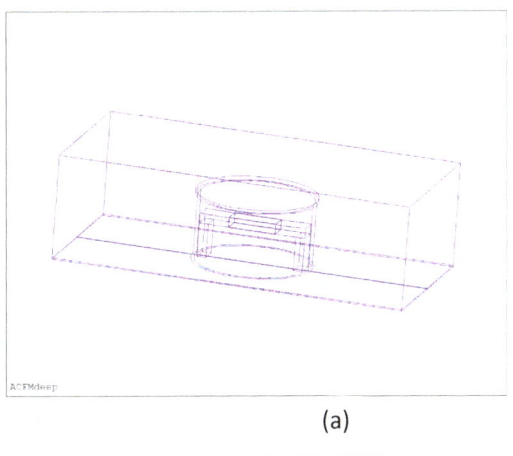

(a)

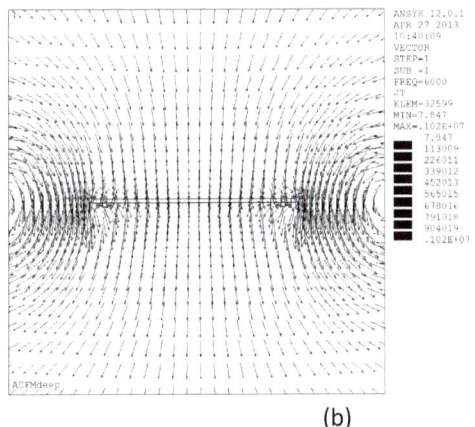

(b)

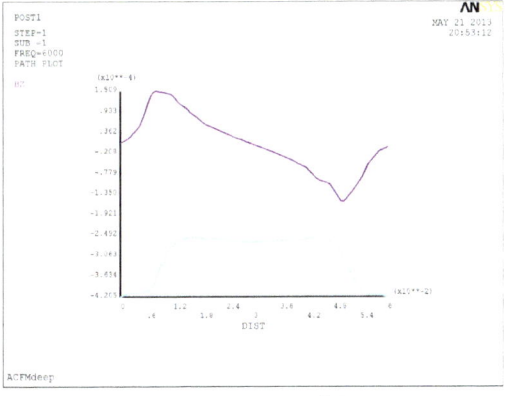

(c)

Table 1 The size of the FEM

Model	Length/mm	Width/mm	Depth/mm
Sample	300	150	10
Core	90.4	16.4	36
Water	300	150	20
Crack	45	0.8	7

Table 2 The characteristic parameters of the FEM

Coil diameter	Number of turns	String material	On-load voltage	Frequency
0.15 mm	500	Steel	1 v	6 k Hz

As shown in Table 3, the ξ_x (40.0%) and ξ_z (36.7%) remain the same basically as the excitation current increases. Hence the sensitivity of detecting signal has no significant effect by the excitation current. However, the larger distortion signal can minimize other interfering signals. The maximum allowable continuous working current of circuit modules is 50 mA for the power constraints. In order to provide strong enough current intensity without the need for cooling the system, the excitation current is set as 40 mA here.

3.3 Fix Structure Lift-Off

Lift-off is the distance between the probe and the specimen; it influences the desired characteristics of the ACFM signals [20]. The magnetic field is weaker as the lift-off distance longer. However, the signals induced and received by ACFM probe are unstable when the lift-off is too small. A small variation of lift-off, which may be due to varying coating thickness, irregularities on sample surface or movement of probe, will lead to a large change in the signal response [21, 22]. As shown in Fig. 4a, b, Bz is an order of magnitude less than Bx. So the weaker Bz component is more susceptible to the lift-off. Simulations are performed to analyze the lift-off effects. Figure 5a shows that the amplitude decreases as the lift-off increases.

As shown in Fig. 5b, the sensitivity of Bz drops steeply as the lift-off increases from 1 to 3 mm. For the uneven specimen or movement of probe, the lift-off value normally changes within ± 1 mm errors. The minimum relative rate of change in signal characteristic vectors is 22.9% with ± 1 mm lift-off variation when the lift-off is below 3 mm. When the lift-off is greater than 4 mm, the maximum relative rate of change is 18.7%. So when the lift-off is below 3 mm, a slight perturbation can result in significant error. Meanwhile, when the lift-off distance is greater than 4 mm, the sensitivity of Bz becomes smooth, which introduce smaller error. However, as the lift-off increases, the amplitude of Bz becomes weak, which is not conducive to detection. To balance the stability and strength of signal, 4 mm is the selected lift-off.

Fig. 4 The optimization for excitation current. **a** The Bx with different excitation current. **b** The Bz with different excitation current. **c** The ΔMax of Bx and Bz against the excitation current

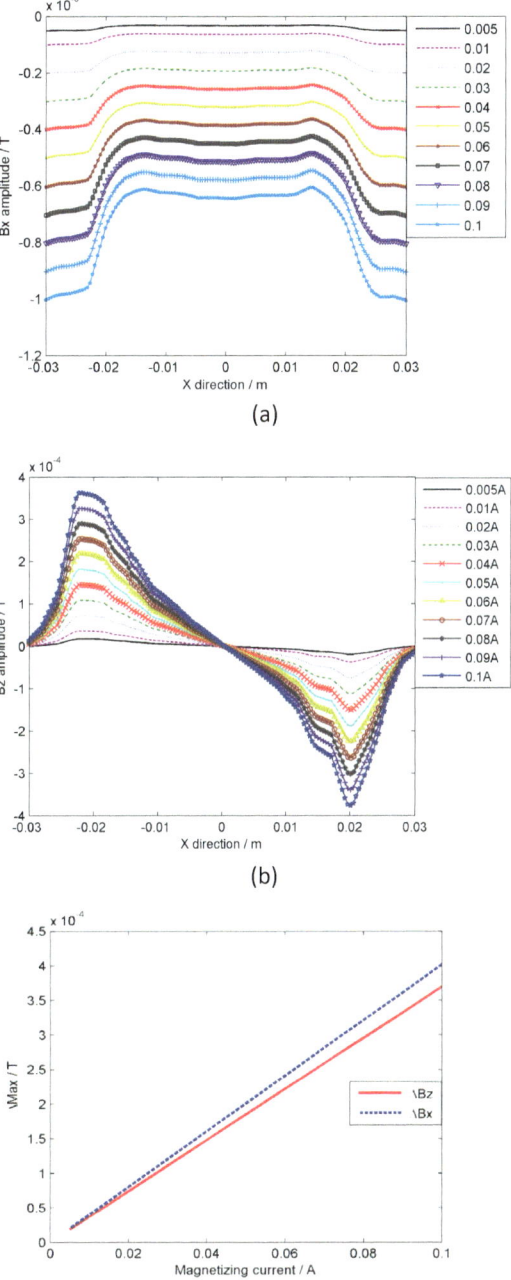

(a)

(b)

(c)

Table 3 The results of simulation with different excitation current

Excitation current	Bx_0	$\Delta Bz_{max}(T)$	ξ_z	$\Delta Bx_{max}(T)$	ξ_x
0.005	−5.03E−05	1.85E−05	0.36742	2.01E−05	0.399594
0.01	−1.00E−04	3.69E−05	0.36743	4.02E−05	0.399642
0.02	−2.01E−04	7.39E−05	0.36743	8.03E−05	0.399572
0.03	−3.02E−04	1.11E−04	0.36741	1.21E−04	0.399609
0.04	−4.02E−04	1.48E−04	0.36742	1.61E−04	0.399587
0.05	−5.03E−04	1.85E−04	0.36742	2.01E−04	0.399594
0.06	−6.03E−04	2.22E−04	0.36742	2.41E−04	0.399599
0.07	−7.04E−04	2.59E−04	0.36743	2.81E−04	0.399602
0.08	−8.04E−04	2.95E−04	0.36742	3.21E−04	0.399600
0.09	−9.05E−04	3.32E−04	0.36742	3.62E−04	0.399591
0.1	−1.01E−03	3.69E−04	0.36743	4.02E−04	0.399642

3.4 Probe Structure

The underwater VACFM probe consists of the excitation coil with a U-shaped ferrite core [23], detecting sensor, excitation circuit, amplifier circuit, conditioning circuit, cover, gland, shell, sealing ring and cable sealing joint, as seen in Fig. 6. These are encapsulated in the probe shell. The material of the shell is 316L (00Cr17Ni14Mo2), which is a kind of non-magnetic stainless steel. The cover of the probe is on the bottom of the shell, which uses non-Magnetic plexiglass. To keep the lift-off, the detecting sensor is fixed on the cover at the thickness of 3 mm (the lift-off of the sensor center is 4 mm). The U-shaped ferrite core, excitation circuit, amplifier circuit and conditioning circuit are fixed in the shell. To seal against the water pressure, the gland compresses the cover with a Sealing ring and all the signal wires pass through the cable sealing joint. The signals are transmitted via signal wires between the underwater and topside.

4 System Performance Testing

4.1 Experimental System

Figure 2 shows the experimental system of underwater VACFM. It consists of two main parts: the underwater probe, the topside signal acquisition and processing system. According to the results of simulations, the experiment parameters are adopted in Table 4.

The excitation coil consists of 500 turns of winding wire on the beam of U-shaped ferrite core with an excitation frequency 6 kHz and magnitude 1 V [24]. The detecting

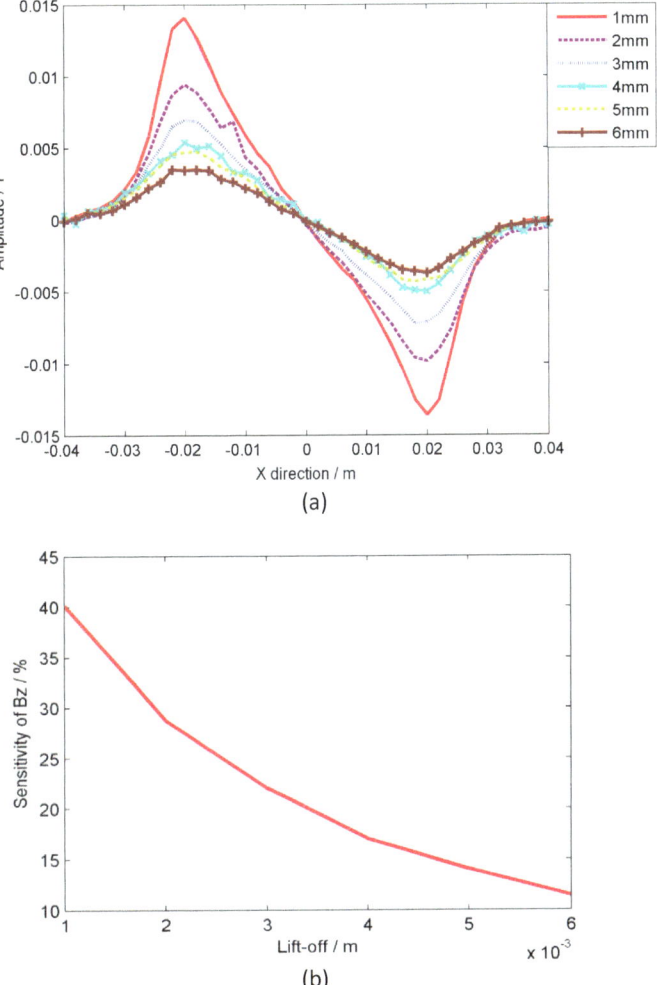

Fig. 5 The optimization for lift-off. **a** The Bz with diffreent lift-off. **b** The sensitivity of Bz against the lift-off

sensor is made up of two coils (the Bx coil is 150 turns and Bz coil is 200 turns) with one common magnetic core. The detecting planes of Bx and Bz coil are perpendicular to X-axis and Z-axis direction respectively.

The test piece is a mild steel specimen with three cracks. The defects are rectangular-shaped cracks with different lengths and same depth, length, which are artificially introduced using the electric discharge machine, as shown in Fig. 7a, the size of cracks are given in Table 5. The 3D scanner is driven by stepper motors and the test system is shown in Fig. 7b. By driving the underwater VACFM probe above the surface of crack (No.1 crack) on the slab at a certain speed in the water tank,

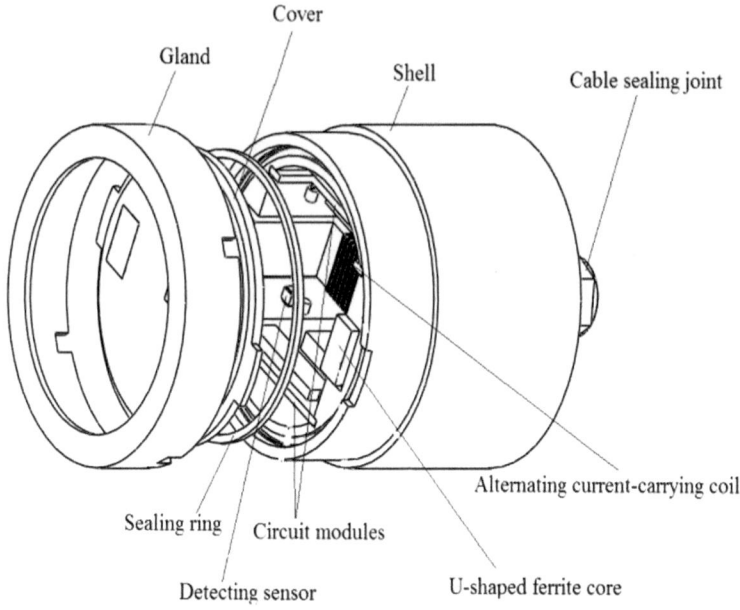

Fig. 6 The structure of underwater VACFM probe

Table 4 The experiment parameters of underwater VACFM

Excitation frequency	Excitation current	On-load voltage	Lift-off
6 kHz	40 mA	1 V	4 mm

the test results are showed on the PC, as shown in Fig. 7c, which includes the Bx and Bz, Butterfly plot and Inverse results. From the Bx and Bz, the crack length can be detected. Sometimes, the curves of Bx and Bz could not be distinguished for nonuniform scanning speed. At the moment, the butterfly plot is useful to identify the crack, in which Bx is plotted against Bz. A loop butterfly plot helps the operator to decide whether a crack is present or not for decreasing misdetection frequency [25]. The Bx signal with scanning speed is processed by the software to calculate the profile of crack. By this inversion operation, the crack profile can also be seen in the inverse results, which is described in previous studies [26]. Thus, the crack can be identified visually by the VACFM.

4.2   Accurate Measurements

Crack lengths are critical parameters for underwater structures safety estimation [27]. In the laboratory, the VACFM probe is drive by stepper motor and screws keeping a

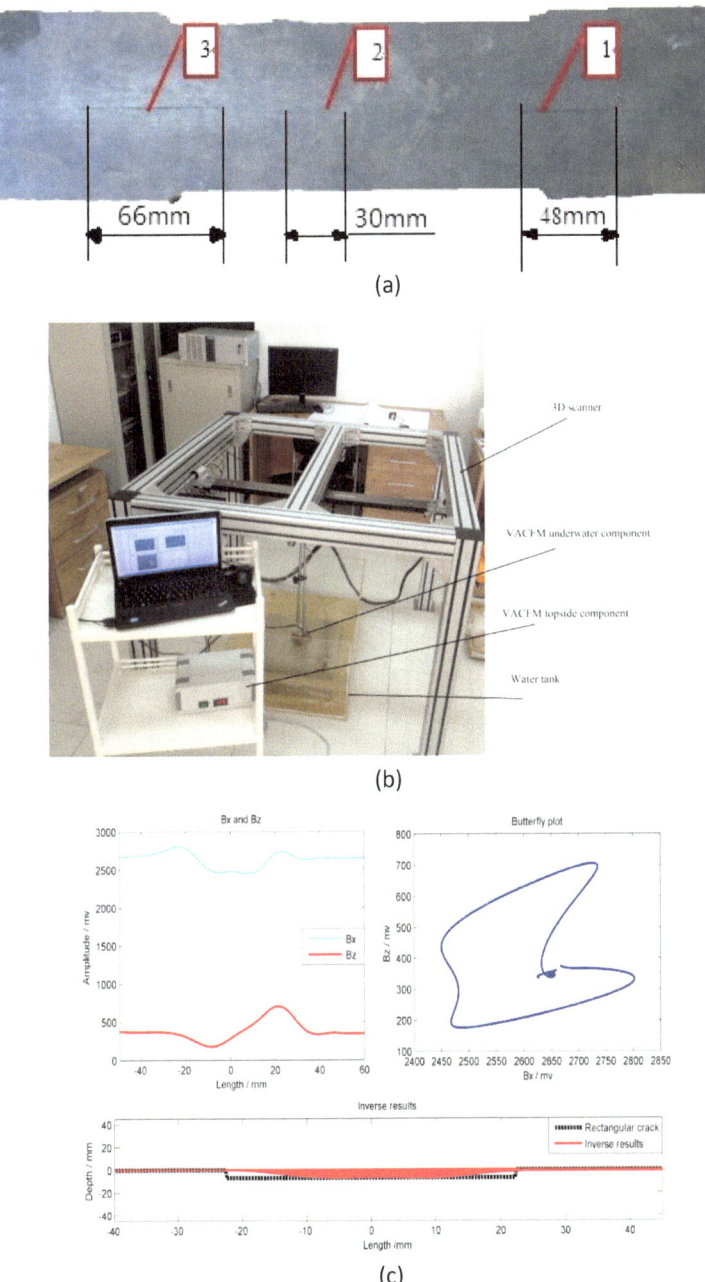

Fig. 7 The specimen and test results. **a** The crack on the surface of the test piece. **b** The VACFM test system. **b** The test results

Table 5 The size of cracks

Crack number	Length (mm)	Width (mm)	Depth (mm)
No. 1	48	0.8	7
No. 2	30	0.8	7
No. 3	66	0.8	7

certain speed. The detecting signals from probe are sent to the topside and collected by A/D card and processed by software. In this process, the signal lags seriously. So there is a great error in the scanning time and the nominal time of the signal. What's more, other interference factors cannot to be neglected for the detection precision of crack length.

To offset the error of the lag of signal processing and other interference factors, TCM is introduced using a compensation factor time-constant ($\bar{t}$). A specimen is calibrated before the test experiment, which has the same material with SUT. With the same material, the same scanning speed, the same interference factors and the same lag degree of signal processing, crack length can be sized accurately with the help of compensation factor $\bar{t}$.

Firstly, the calibration crack is detected for several times at a certain speed to get a stable $\bar{t}$. The lengths of the crack can be given as follows:

$$L_1 = t_1 V N_1$$
$$L_1 = t_2 V N_2$$
$$\cdots\cdots\cdots$$
$$L_1 = t_6 V N_6 \tag{3}$$

L_1 is actual length (AL) of crack. V is scanning speed. N is the number of sampling points (NSP) between the peak and trough of crack, t is the time-constant. So the mean value of t can be given as:

$$\bar{t} = \frac{L_1}{6V}\left(\frac{1}{N_1} + \frac{1}{N_2} + \cdots + \frac{1}{N_6}\right) \tag{4}$$

The scanning speed is calculated by the rotate speed of stepper motor and the transmission ratio of screws. So the scanning speed should be chosen at some special value only. The target crack is scanned at a certain speed and the measuring length (*ML*) of the crack is given as follows:

$$L = \bar{t} V N \tag{5}$$

To get $\bar{t}$, No.1 crack (48 mm length) is tested for 6 times at 3.81 mm/s (select a certain speed). The NSP between the peak and trough of the crack is shown in Table 6.

Table 6 The NSP between the peak and trough of No.1 crack

Scan times	1	2	3	4	4	6
NSP	978	1195	1002	1031	893	1101

Table 7 The test results of No. 2 and No. 3 crack at 3.81 mm/s

Crack number	V (mm/s)	NSP	ML (mm)	Relative error
2	3.81	702	32.87	9.56%
3	3.81	1395	65.32	1.03%

Table 8 The test results of No. 2 and No. 3 crack at 4.52 and 5.08 mm/s respectively

Crack number	V (mm/s)	$\bar{t}$ (s)	ML (mm)	Relative error (%)
2	4.52	13.16E−3	34.01	13.37
	5.08	14.45E−3	36.22	20.73
3	4.52	13.16E−3	67.54	2.33
	5.08	14.45E−3	71.53	8.38

The $\bar{t}$ can be obtained from (4), $\bar{t} = 12.29\text{E} - 3$ s. The No.2 (30 mm length crack) and No.3 (66 mm length crack) are also tested at 3.81 mm/s. According to (5), the length of No.2 and No.3 crack can be obtained, the test results and relative errors ($|AL - ML|/AL$) are shown in Table 7.

To analyze the effect of scanning speed on detection precision, No.1 crack (calibration crack) is tested at 4.52 and 5.08 mm/s and the No.2 and No.3 crack (target crack) are tested at the same speed respectively according to the TCM. The test results are showed in Table 8 and the relationships between the relative errors and the scanning speeds are displayed in Fig. 8.

4.3 Results and Discussion

Comparing the results of the experiments and simulations, as seen in Figs. 3 and 7, a good agreement is shown in trends of Bx and Bz. The troughs of Bx (the absolute value of Bx presents a trough in the simulation result) lie in the center of the crack, meanwhile the peaks and troughs of Bz locates at the both ends of the crack.

As shown in Table 8, the relative error of the crack length increases as the scanning speed increase using the VACFM based on TCM. Meanwhile, comparing the results of No.2 and No.3 crack, the relative error of No.3 crack (66 mm length) is much less than that of No.2 crack (30 mm length). It shows that the VACFM is more accurate to longer crack. It means that, to get an optimum test results, the scanning speed should be slower. Of course, low efficiency of scanning speed will make it inadequacy in

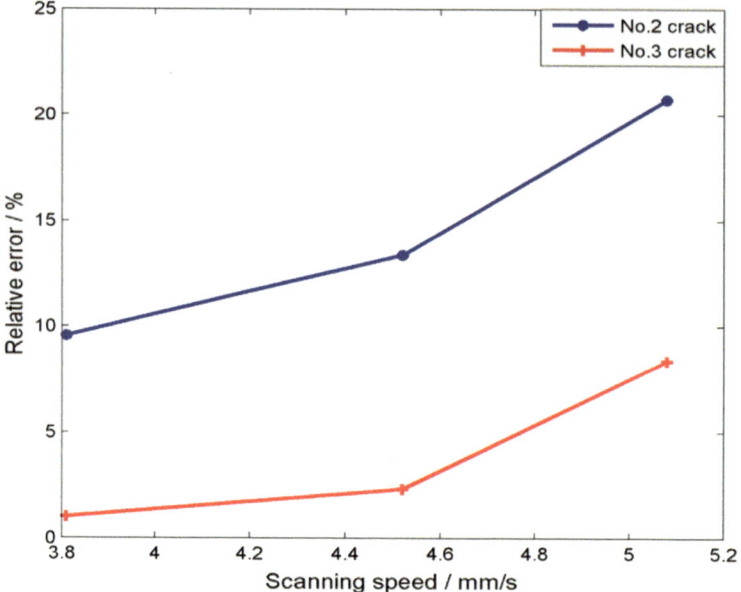

Fig. 8 The relative error against the scanning speed

service. To meet the needs of detection precision in engineering, whose relative error should be less than 10%, the 3.81 mm/s can be selected as the optimum scanning speed.

5 Design Summary and Conclusion

In this work, the VACFM system has been built including the excitation coil, detecting sensor, probe structure, signal processing system and software system. To reduce the signal attenuation in long-distance transmission, the excitation and signal processing circuits (amplification and filtering) are encapsulated the underwater probe. The parameters (such as excitation current, lift-off and structure) of VACFM system are optimized by the modeling to achieve better sensitivity, stability and accuracy. Meanwhile, TCM is presented to improve the detection precision by reducing the effect of signal attenuation, interference factor and the lag of signal processing. According to TCM, the scanning speed is optimized.

The modeling and optimization of visual ACFM system are able to provide a high accurate and visual measurement for underwater structure cracks, which is tested and proved by real cracks inspection experiments. The results show that the underwater VACFM based on TCM can size the crack length and profiling the section of crack with the optimized system, which meets the needs of detect precision in engineering.

In a word, the underwater VACFM system provides a method for the intelligent and accurate detection of underwater structure cracks to keep subsea oil and gas equipment safety and reliability. The test results of the system are visual, which is convenient and simple for the inspector to evaluate the cracks. Furthermore, the underwater VACFM system presented this paper are applicable to other long-distance transmission test fields.

The performance of VACFM is still not robust for detection of complicated shape underwater defects. Moreover, further research needs to focus on detecting and reconstructing different type of underwater cracks and improving detection precision.

References

1. M.F. Randolphn, C. Gaudin, S.M. Gourvenec, D.J. White, N. Boylan, M.J. Cassidy, Recent advances in offshore geotechnics for deep water oil and gas developments. Ocean Eng. **38**(7), 818–834 (2011)
2. A. Gholizad, A.A. Golafshani, V. Akrami, Structural reliability of offshore platforms considering fatigue damage and different failure scenarios. Ocean Eng. **46**, 1–8 (2012)
3. P. Traverso, E. Canepa, A review of studies on corrosion of metals and alloys in deep-sea environment. Ocean Eng. **87**, 10–15 (2014)
4. N. Hifi, N. Barltrop, Correction of prediction model output for structural designand risk-based inspection and maintenance planning. Ocean Eng. **97**, 114–125 (2015)
5. W.D. Dover, R. Collins, D.H. Michael, The use of AC-field measurements for crack detection and sizing in air and underwater. J. Philos. Trans. R. Soc. B **320**(1554), 271–283 (1986)
6. D. Naso, B. Turchiano, P. Pantaleo, A fuzzy-logic based optical sensor for online weld defect-detection. IEEE Trans. Ind. Inf. **1**(4), 259–273 (2005)
7. D.M. Syahkal, R.F. Mostafavi, 1–D probe array for ACFM inspection of large metal plates. J. IEEE Trans. Instrum. Meas. **51**(2), 374–382 (2002)
8. A. Noroozi, R.P.R. Hasanzadeh, M. Ravan, A fuzzy learning approach for identification of arbitrary crack profiles using ACFM technique. J. IEEE Trans. Magn. **49**(4), 5016–5027 (2013)
9. M. Smith, R. Sutherby, The detection of pipeline SCC flaws using the ACFM technique. Insight. J. **47**(12), 765–768 (2005)
10. R.A. Esmaeel, J. Briand, F. Taheri, Computational simulation and experimental verification of a new vibration-based structural health monitoring approach using piezoelectric sensors. J. Struct. Health Monit. **11**(2), 1663–1676 (2012)
11. H.P. Tan, R. Diamant, W.K.G. Seah, M. Waldmeyer, A survey of techniques and challenges in underwater localization. Ocean Eng. **38**, 818–834 (2011)
12. S.L. Dai, C. Wang, F. Luo, Identification and learning control of ocean surface ship using neural networks. J. IEEE Trans. Ind. Informat. **8**(4), 801–810 (2012)
13. Y.C. Cao, W.W. Yu, W. Ren, G.R. Chen, An overview of recent progress in the study of distributed multi-agent coordination. IEEE Trans. Ind. Informat. J. **9**(1), 427–438 (2013)
14. V. Pallayil, M. Chitre, S. Kuselan, A. Raichur, M. Ignatius, J.R. Potter, Development of a second-generation underwater acoustic ambient noise imaging camera. IEEE J. Ocean. Eng. **99**, 1–5 (2015)
15. M. Lugg, The first 20 years of the A.C. field measurement technique, in *Proceedings of the 17th World Conference on Nondestructive Testing* (Shanghai, 2008), pp. 25–28
16. X.L. Bai, Y.M. Fang, W.S. Lin, L.P. Wang, B.F. Ju, Saliency-based Defect detection in industrial images by using phase spectrum. J. IEEE Trans. Ind. Inf. **10**, 2135–2145 (2014)
17. K. Li, G.Y. Tian, L. Cheng, A. Yin, W. Cao, S. Crichton, State detection of bond wires in IGBT modules using eddy current pulsed thermography. J. IEEE Trans. Power Elec. **29**(9), 5000–5009 (2014)

18. G.L. Nicholson, C.L. Davis, Modelling of the response of an ACFM sensor to rail and rail wheel RCF cracks. J. NDTE Int. **46**, 107–114 (2012)
19. Y. Gotoh, K. Sakurai, N. Takahashi, Electromagnetic inspection method of outer side defect on small and thick steel tube using both AC and DC magnetic fields. J. IEEE Trans. Magn. **45**(10), 4467–4470 (2009)
20. W. Li, G.M. Chen, X.K. Yin, C.R. Zhang, T. Liu, Analysis of the lift-off effect of a U-shaped ACFM system. NDTE Int. **53**, 31–35 (2013)
21. A.H. Salemi, S.H.H. Sadeghi, R. Moini, The effect of magnetic field sensor lift-off on SMFM crack signals, in *Proceedings of the QNDE* (New York, 2001), pp. 977–983
22. G.M. Chen, W. Li, Z.X. Wang, Structural optimazation of 2-D array probe for alternating current field measurement. J. NDTE Int. **40**(6), 455–461 (2007)
23. H. Hoshikawa, K. Koyama, Basic study of a new ECT Probe using uniform rotating direction eddy current, in *Proceedings of the QNDE* (1997), pp. 1067–1074
24. W. Li, G.M. Chen, W.Y. Li, Z. Li, F. Liu, Analysis of the excitation frequency of a U-shaped ACFM system. J. NDTE Int. **44**, 324–328 (2012)
25. M.C. Lugg, The first 20 years of the A.C. field measurement Technique, in *WCNDT* (2012), pp. 16–20
26. W. Li, G.M. Chen, Defect visualization for alternating current field measurement based on the double u-shape inducer array. J. JME. **45**, 233–237 (2009)
27. Y. Fedala, M. Streza, F. Sepulveda, J.-P. Roger, G. Tessier, C. Boué, Infrared lock-in thermography crack localization on metallic surfaces for industrial diagnosis. J. Nondestruct. Eval. **33**(3), 335–341 (2014)

Research on High-Precision Evaluation of Crack Dimensions and Profiles Methods for Underwater Structure Based on ACFM Technique

Abstract The conventional alternating current field measurement (ACFM) technique evaluates the crack dimensions by the distance between Bz peaks and the Bx trough. The interaction of the crack length and depth on the characteristic signals is not considered before, which results in the low accuracy evaluation results of crack dimensions and profiles. In the light of the problems above, the 3D simulation model of ACFM is set up in the underwater environment. The influence rules of the crack length and depth on the characteristic signals are analyzed. On this basis, the two-step interpolation crack dimensions evaluation algorithm and the segmentation interpolation crack profiles reconstruction algorithm are presented to achieve high-precision evaluation of crack dimensions and profiles. The underwater ACFM testing system is set up and the crack detection experiments are carried out. The results show that the crack depth does not affect the distance between the peak and trough of the Bz. The crack length that is shorter than 30 mm affects the trough of the Bx. The two-step interpolation algorithm based on the Bx and Bz can achieve high-precision evaluation of the crack length and depth. The maximum errors of the evaluated length and depth are 3.0 and 7.5% respectively. The segmentation interpolation algorithm based on the Bx can achieve high-precision visual reconstruction of the crack profile. The maximum error of the reconstructed profile is 5.7%.

Keywords ACFM · Crack · Dimension and profile · High-precision · Interpolation algorithm · Underwater structure

1 Introduction

Alternating Current Field Measurement (ACFM) technology is a newly emerging electromagnetic non-destructive testing technology in recent years. It has many advantages such as lift-off insensitivity, non-contact detection, no need to clean structural attachments and coatings, and quantitative detection. It has been widely used in underwater structures, high-speed rail, nuclear power and non-destructive testing of surface defects of special equipment [1–3]. The alternating current magnetic field

© The Author(s) 2024
X. Yuan et al., *Recent Development of Alternating Current Field Measurement Combine with New Technology*, https://doi.org/10.1007/978-981-97-4224-0_8

relies on a rectangular excitation coil to induce a uniform current on the surface of the conductive test block. The induced current passes through the crack vertically, gathers at both ends of the crack, causing the magnetic field (Bz) perpendicular to the direction of the test block to generate positive and negative peaks at both ends of the crack. Therefore, the distance between the positive and negative peaks of Bz reflects the crack length information; the induced current alternately generates peaks and valleys in the magnetic field (By) perpendicular to the direction of the crack at the end of the crack [4–6]; the induced current bypasses the bottom of the crack center, the current density at the center of the crack weakens, causing the magnetic field (Bx) along the direction of the crack to generate a valley [7], the valley reflects the crack depth information, as shown in Fig. 1.

Under normal circumstances, the crack length can be obtained based on the distance between the peaks and valleys of the characteristic signal Bz, and the approximate depth of the crack can be obtained based on the change in the valley of the characteristic signal Bx. However, due to the large error between the crack length and the distance between the peaks and valleys of the characteristic signal Bz, and the impact of the crack length on the Bx valley, the traditional characteristic signal evaluation method has a large error, affecting the evaluation of the remaining life of the structure and maintenance decisions. W. D. Dover, A. M. Lewi, etc. established the classic ACFM theoretical model, and gave an analytical model of the electromagnetic field disturbance around the crack based on the two-dimensional plane assumption and uniform induced current [8, 9]. Hu Shuhui and others proposed a linear interpolation method to invert the crack length and depth dimensions [10]. A.

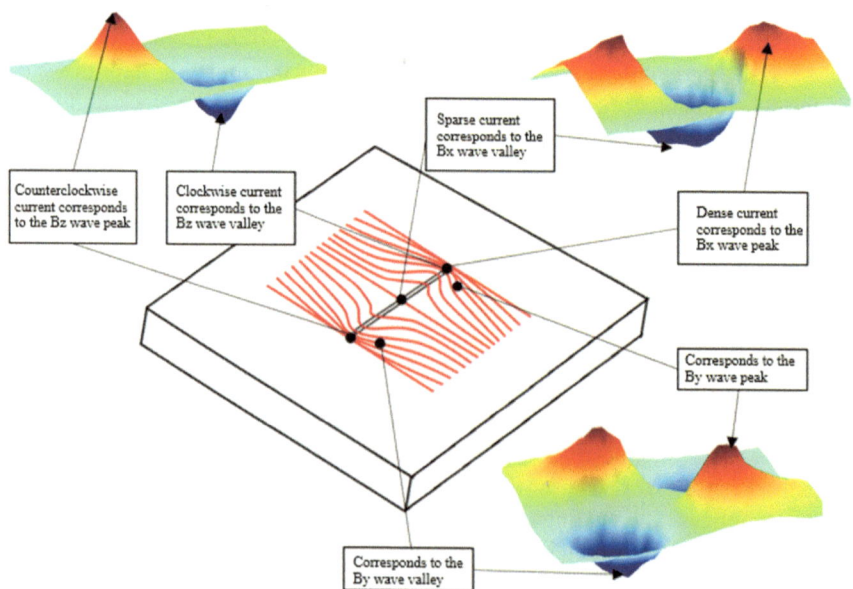

Fig. 1 Principle of ACFM

L. Ribeiro and others gave a forward model of the crack and characteristic signal under uniform induced current [11]. G. L. Nicholson and others applied ACFM to the detection and evaluation of cluster rolling contact fatigue (RCF) cracks in rails [12, 13]. R. K. Amineh [14] and others proposed a crack depth evaluation method under different lift-off heights based on the model inversion algorithm. The research group and other scholars have proposed a crack size inversion algorithm based on neural network self-learning in previous studies [15–18]. The above research uses the characteristic signal Bx to analyze the crack depth, and uses the characteristic signal Bz to analyze the crack length, laying the foundation for the ACFM detection and evaluation of the crack. However, the above studies did not consider the interaction between the crack length and depth on the characteristic signals Bx and Bz, resulting in insufficient evaluation accuracy. In addition, self-learning methods such as neural networks require a large amount of sample data, which is difficult to apply and implement quickly on site.

In response to the above problems, this paper establishes a three-dimensional finite element simulation model of ACFM in a seawater environment, analyzes in detail the interaction rules of the changes in crack length and depth on characteristic signals, and proposes a two-step interpolation crack size evaluation algorithm and a segmented interpolation crack profile reconstruction algorithm. This provides an effective method for real-time detection and high-precision evaluation of cracks in underwater structures.

2   Marine Environment ACFM Simulation Model

A three-dimensional simulation model of ACFM in a seawater environment is established using the ANSYS finite element simulation software [19, 20]. As shown in Fig. 2, the simulation model mainly consists of an excitation coil, a U-shaped manganese-zinc ferrite magnetic core, a steel plate test block, and a crack defect. The excitation coil is wound on the manganese-zinc ferrite crossbeam, and an alternating current of 0.1 A with a frequency of 2000 Hz is loaded on the excitation coil. To realistically simulate the distribution of the electromagnetic field in the seawater environment, the inside of the simulation model probe is filled with air, and the outside is surrounded by the seawater environment. The characteristic parameters of the simulation model are shown in Table 1, and the structural parameters are shown in Table 2.

The current vector diagram on the surface of the steel plate test block in the simulation model is extracted, as shown in Fig. 3a. It can be seen that the excitation coil induces a uniform current on the surface of the test block, and the current passes through the crack vertically, gathering at both ends and deflecting in opposite directions. The current density on the surface of the steel plate test block is extracted, as shown in Fig. 3b. The current density gathers at the end of the crack to form a maximum peak, and the induced current bypasses from the bottom of the crack to form a valley in the center of the crack, causing Bx to appear a valley at the center

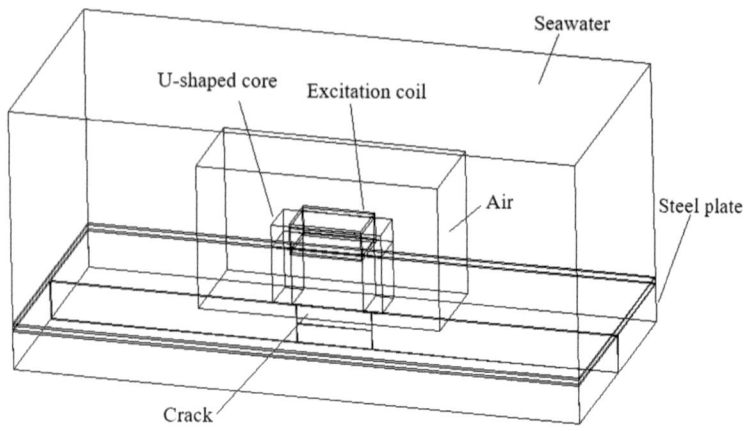

Fig. 2 3D FEM model of ACFM in seawater environment

Table 1 Parameters of simulation model

Coil diameter /mm	Coil number of turns/ circle	Load current/ A	Frequency/ Hz	Steel plate phase to permeability	Steel plate electricity resistivity/ Ω·m	Seawater hydropower resistivity/ Ω·m	Core phase to permeability
0.15	500	0.1	2000	1000	9.78E−8	0.22	10,000

Table 2 Dimensions of simulation model

Project	Length/mm	Width/mm	Height/mm	Core leg length/mm	Core horizontal height/mm	Coil thickness/ mm
Core	29	11	21	5	6	–
Workpiece	150	70	10	–	–	–
Coil	19	–	–	–	–	1
Housing	64	26	38	–	–	–
Crack	30	0.5	5	–	–	–

of the crack, as shown in Fig. 4a. Due to the different deflection directions, the disturbance current causes Bz to present positive and negative peaks at the position of the crack end, as shown in Fig. 4b. The change rules of the characteristic signals Bx and Bz in the simulation model in the seawater environment are consistent with the ACFM principle.

The characteristic signal Bx is related to the crack depth because the current bypasses in the direction of the crack depth, and the peaks and valleys of the characteristic signal Bz are located at the two ends of the crack, so the distance between the

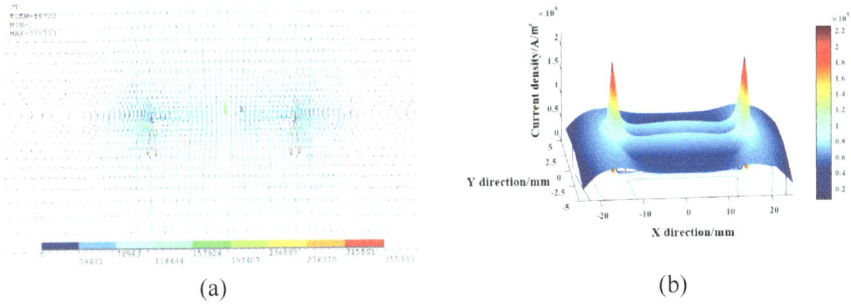

Fig. 3 Distribution of induced current in the surface of specimen. **a** Current vector diagram. **b** Current density plot

peaks and valleys reflects the crack length. However, the changes in crack length and depth have an interactive effect on the characteristic signals Bz and Bx, and a single change in length or depth is not the only factor affecting the characteristic signals. In order to determine the influence of crack size changes on the characteristic signals Bx and Bz, the characteristic signals of different size cracks are analyzed with the help of simulation.

3 Two-Step Interpolation Algorithm

Four groups of simulation models with different sizes of cracks were established, all with a crack width of 0.5 m. The first group of simulation models are cracks of the same length (10 mm) with different depths (1, 2, 3, 4, 5, 6 mm). The second group is cracks of the same length (20 mm) with different depths (1, 2, 3, 4, 5, 6 mm). The third group is cracks of the same length (30 mm) with different depths (1, 2, 3, 4, 5, 6 mm). The fourth group is cracks of the same length (40 mm) with different depths (1, 2, 3, 4, 5, 6 mm). The fifth group is cracks of the same length (50 mm) with different depths (1, 2, 3, 4, 5, 6 mm). The distance between the positive and negative peaks of the characteristic signal Bz is defined as P_{Bz}, and the P_{Bz} change graph of cracks with different depths is obtained, as shown in Fig. 5a. The change rule of P_{Bz} of the characteristic signal with different depths is basically consistent with the crack length, indicating that P_{Bz} is basically not affected by the crack depth.

In order to compare the distortion amplitude of the characteristic signals of cracks at different depths, and to eliminate the influence of linear parameters such as current and coil turns on the simulation or experimental results, the sensitivity of the characteristic signal Bx is defined as S_{Bx}.

$$SBx = (Bx_0 - Bx_{min})Bx_0 = 1 - Bx_{min}/Bx_0 \qquad (1)$$

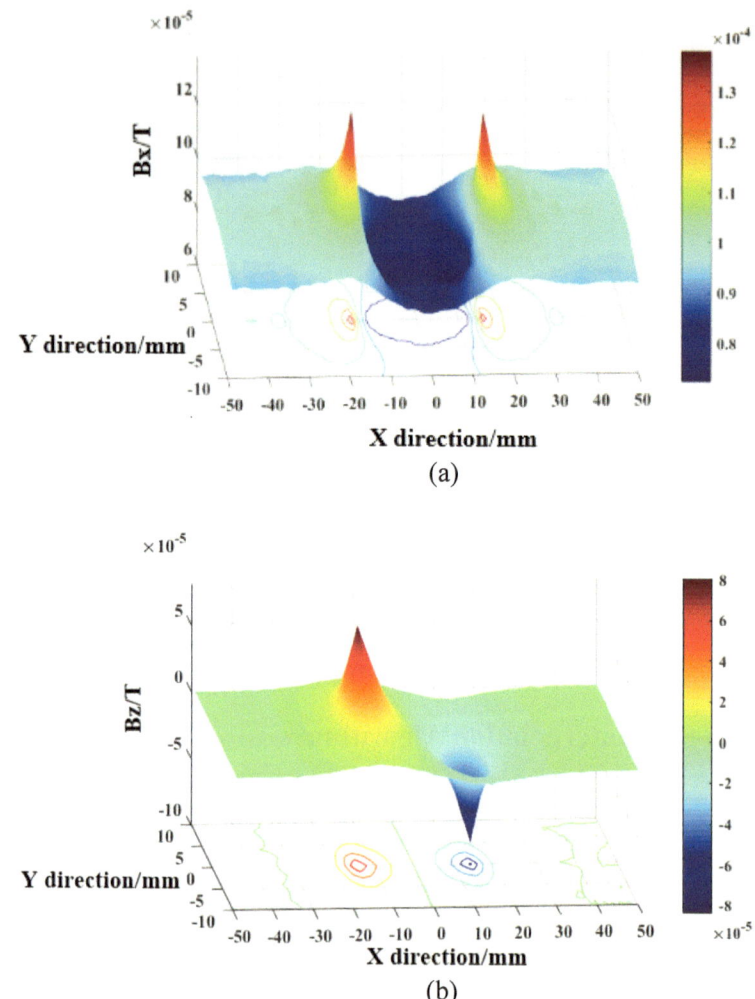

Fig. 4 Distorted magnetic field. **a** Bx. **b** Bz

The sensitivity S_{Bx} of the maximum value position distortion of the characteristic signal of cracks of different lengths is obtained, as shown in Fig. 5b. The characteristic signal Bx is mainly the secondary magnetic field distortion caused by the current disturbance in the center of the crack, and the Bx disturbance situation is more obviously affected by the depth of the crack, and the Bx sensitivity of the characteristic signal of cracks of different depths differs greatly. At the same time, the length of the crack also affects the disturbance of the current in the depth direction. The current in the center of the shorter crack is affected by the length of the crack, causing the maximum sensitivity change of Bx. When the crack extends to a certain extent, the induced current in the center of the crack reaches a minimum and is not

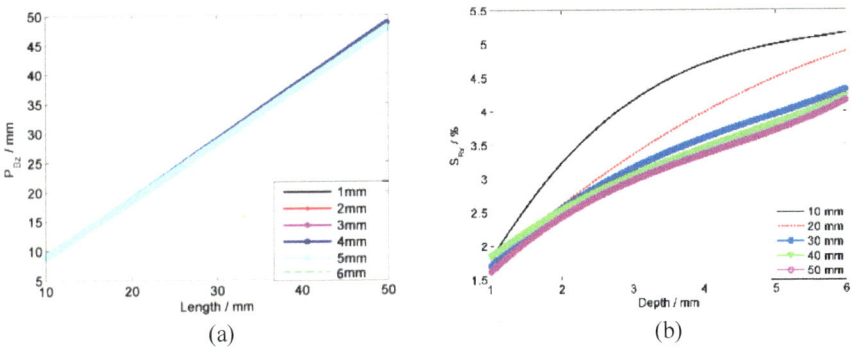

Fig. 5 Signal characteristic with crack size. **a** PBz with cracks of different depths. **b** SBx with cracks of different lengths

affected by the length of the crack, and the Bx sensitivity of the characteristic signal is also not affected by the length of the crack. As can be seen from Fig. 5b, when the crack length is greater than 30 mm, the Bx sensitivity of the characteristic signal of cracks of different lengths basically remains consistent. When the crack length is less than 30 mm, there is a large difference in the sensitivity of the characteristic signal of cracks of different lengths.

In summary, the peak-to-valley distance of the characteristic signal Bz is affected by the crack length and is not affected by the change in crack depth; the sensitivity of the characteristic signal Bx is affected by the crack depth and is also affected by the crack length. Based on the above rules, in order to achieve high-precision evaluation of cracks, a two-step interpolation algorithm based on characteristic signals Bx and Bz is proposed, with the following specific steps:

(1) Use the characteristic signal Bz to obtain the peak-to-valley distance P_{Bz}, and use PBz to obtain the crack length L.

(2) When the crack length L $\geq$ 30 mm, the sensitivity S_{Bx} and the crack depth interpolation formula are obtained by simulation or experimental fitting, and the crack depth is evaluated by the interpolation method measured by the experiment S_{Bx}. When the crack length L < 30 mm, the sensitivity S_{Bx} and the crack depth interpolation formula of a specific length (such as 10, 20 mm, etc.) crack are obtained by simulation or experimental fitting, and the crack depth is evaluated by selecting the approximate length fitting formula measured by the experiment S_{Bx}. The two-step interpolation method can quickly obtain crack length and depth information, does not require a large amount of sample data, and is conducive to the online real-time evaluation of crack size.

The crack length, depth and characteristic signal rules obtained from the simulation model in this article, and the formulas (2)–(4) obtained from polynomial interpolation, where formula (2) is the first step to obtain the crack length L, formula (3) is the second step to obtain the crack depth D_{30} greater than 30 mm in length, and formula (4) is the second step to obtain the crack depth D_{20} of 20 mm in length.

$$L = 1.02 \times P_{Bz} + 1.02 \tag{2}$$

$$D_{30} = -0.2106 \times S_{Bx}^3 + 2.207 \times S_{Bx}^2 - 5.171 \times S_{Bx} + 4.491 \tag{3}$$

$$D_{20} = 0.07196 \times S_{Bx}^3 - 0.4911 \times S_{Bx}^2 + 2.296 \times S_{Bx} - 1.908 \tag{4}$$

4 Establishment of the Experimental System

The underwater ACFM detection system consists of an underwater probe, a hull, and an above-water computer. The probe is connected to the hull through a water-sealed joint, and the hull is connected to the above-water computer through an optical fiber, as shown in Fig. 6a. The internal lithium battery of the hull powers the entire system. The excitation module generates a sinusoidal excitation signal with a frequency of 2000 Hz and an amplitude of 10 V and loads it into the excitation coil in the probe. The excitation coil induces a uniform current field on the surface of the test piece. When a defect is present, the induced current is disturbed, causing spatial magnetic field distortion. The magnetic field sensor inside the probe measures the distorted magnetic field and transmits it to the amplification module inside the hull through the primary amplification filter circuit [21, 22]. After the signal is amplified, it is transmitted to the acquisition card, transmitted to the processor after A/D conversion, and transmitted to the surface through the optical fiber. The optical fiber receiver on the surface converts the optical signal into an electrical signal and transmits it to the computer. The internal program of the computer analyzes and evaluates the crack size in real time. The developed underwater ACFM system is shown in Fig. 6b.

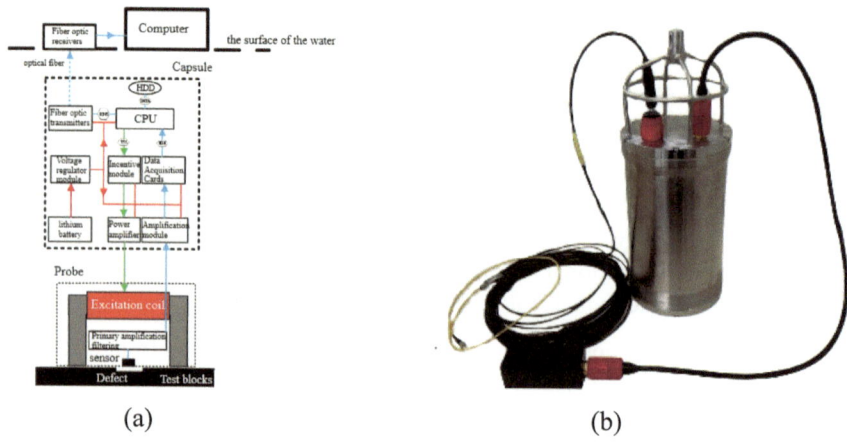

(a)   (b)

Fig. 6 Underwater ACFM system. **a** System block diagram. **b** System photo

5 Crack Evaluation Test

The probe and hull are placed in a water tank filled with seawater medium. The probe is driven by a mechanical arm to scan the crack area of the test block at a uniform speed of 40 mm/s, as shown in Fig. 7a. Two test blocks are set up in this test, both of which are made of Q235 material. Test block 1 has rectangular groove cracks of the same depth (4 mm), with a crack opening of 0.5 mm and crack lengths of 20, 40, and 45 mm respectively, as shown in Fig. 7b. Test block 2 has cracks of different cross-sectional shapes, with a crack opening of 0.5 mm. Crack 1# is a semi-elliptical crack with a length of 20 mm and a maximum defect depth of 4 mm; Crack 2# has a semi-elliptical defect with a length of 30 mm and a maximum defect depth of 5 mm; Crack 3# has a complex shape with a surface opening length of 40 mm and a maximum defect depth of 4 mm, as shown in Fig. 7c.

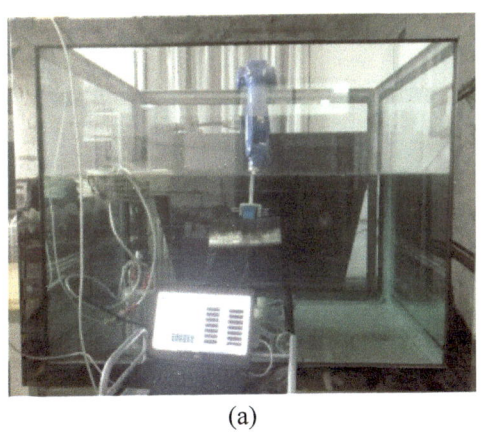

(a)

cracks of different lengths (depth 4 mm and width 0.5 mm)

Length 45mm   Length 40mm   Length 20mm

(b)

1#   2#   3#

(c)

Fig. 7 Underwater ACFM testing system. **a** Test photos. **b** Test block 1. **c** Specimen block 2 sections

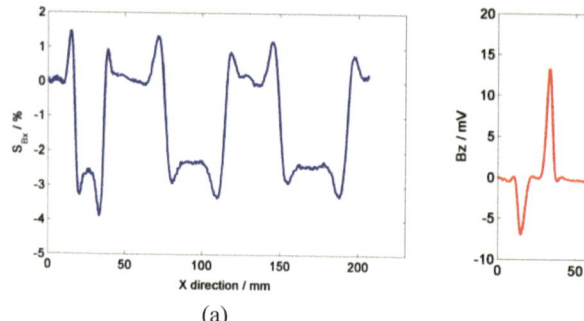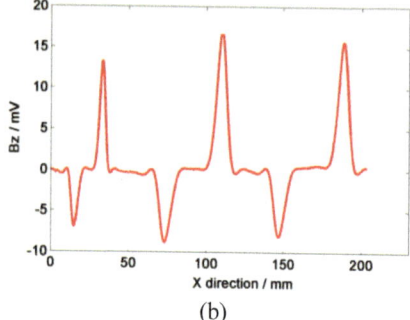

(a)   (b)

Fig. 8 Testing results of cracks with same length. **a** S_{Bx}. **b** Bz

5.1 Crack Size Evaluation

The probe is driven by the mechanical arm to uniformly detect Test Block 1. The detection results of the sensitivity SBx of the characteristic signal Bx are shown in Fig. 8a, and the detection results of the characteristic signal Bz are shown in Fig. 8b.

From Bz, the peak-to-valley distances PBz can be obtained as 18.0, 37.5, and 42.0 mm, respectively. Using formula (2) for interpolation, the crack length sizes are obtained as 19.4, 39.3, and 43.9 mm. For cracks with lengths of 39.3 and 43.9 mm, the crack depths are calculated using formula (3) to be 3.7 and 3.8 mm, respectively, with errors of 7.5 and 5.0%, respectively, achieving high evaluation accuracy.

For cracks with a length less than 30 mm, the approximate formula (4) is used to evaluate the crack with a length of 19.4 mm, and the estimated crack depth is 3.8 mm, with an error of 7.5%. If formula (3) is used directly to estimate the crack depth, the crack depth is obtained as 7.2 mm, which is much larger than the actual size of the crack. It can be seen that the two-step interpolation algorithm can improve the evaluation accuracy of the crack depth for cracks with a length less than 30 mm. The results of the evaluation of the length and depth of the crack using the two-step interpolation algorithm are shown in Table 3. In summary, the two-step interpolation algorithm can achieve high-precision evaluation of crack length and depth, with a maximum depth error of 7.5% and a maximum length error of 3.0%.

5.2 Crack Profile Evaluation

The probe uniformly detects Test Block 2, and the sensitivity S_{Bx} of the characteristic signal Bx is obtained by formula (1), and the part greater than the background magnetic field (the signal above the profile is removed), as shown in Fig. 9a.

In order to achieve crack profile reconstruction and evaluation, this paper proposes a crack profile reconstruction algorithm based on the segmentation interpolation of

Table 3 Evaluation results of cracks

Crack length/mm	20	40	45
Crack depth/mm	4	4	4
P_{Bz}/mm	18.0	37.5	42.0
S_{Bx}/%	3.91	3.23	3.28
Length assessment/mm	19.4	39.3	43.9
Length error/%	3.0%	1.8%	2.4%
In-depth assessment/mm	3.8	3.7	3.8
Depth error/%	5.0%	7.5%	5.0%

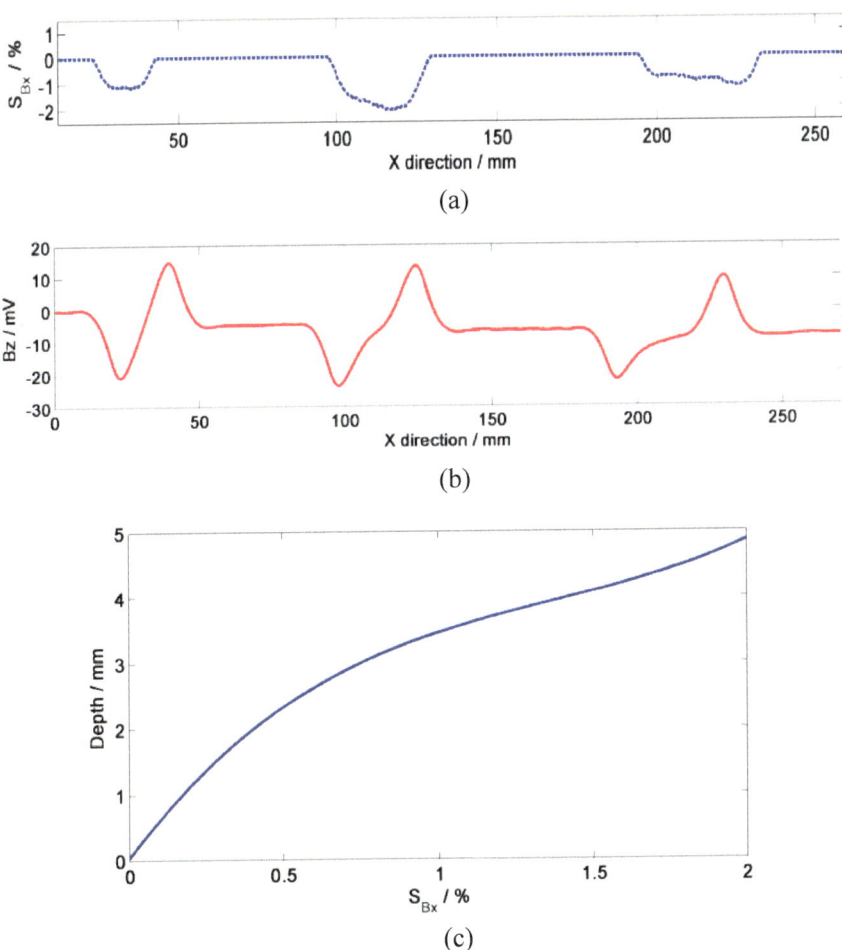

Fig. 9   Testing results of crack with different profiles. **a** S_{Bx}. **b** Bz. **c** Depth and S_{Bx} relationship

the characteristic signal Bx. First, use the peak-to-valley distance P_{Bz} of the characteristic signal Bz in Fig. 9b and formula (2) to revise the crack length, and obtain the lengths of the three cracks as 19.4, 29.6, and 38.8 mm. Secondly, using the segmentation interpolation method, Crack 2# is used as the calibration crack, which is divided into 15 equal parts in the length direction to form 15 crack depth points. At the same time, the sensitivity S_{Bx} of the characteristic signal of Crack 2# is set to 15 equal parts in the crack area to form the sensitivity of 15 position points. Then, the corresponding relationship between the depth of any position of the crack and the sensitivity S_{Bx} can be obtained, as shown in Fig. 9c. The relationship between crack depth D and sensitivity S_{Bx} can be expressed by the polynomial interpolation formula as follows:

$$D = 0.8762 \times S_{Bx}^3 - 3.634 \times S_{Bx}^2 + 6.174 \times S_{Bx} + 0.04038 \tag{5}$$

Finally, the sensitivity S_{Bx} of Crack 1# is divided into 10 equal parts along the length direction to obtain the sensitivity of 10 position points. The sensitivity S_{Bx} of Crack 3# is divided into 20 equal parts along the length direction to obtain the sensitivity of 20 position points. Using formula (5), the crack depth corresponding to the sensitivity of each position point can be obtained, and the crack profile contour can be reconstructed, as shown in Fig. 10.

The vertical coordinate 0 position in Fig. 10a and c represents the upper surface of the test block, and the negative value represents the depth below the surface of the test block. Figure 10b and d show the visualized shape of the crack profile, which matches the defect profile height in Test Block 2. The actual profile area of Crack 1# is 60.76 mm^2, and the area of the closed area that can be evaluated by the curve is 57.28 mm^2, with an error of 5.7%; the actual profile area of Crack 3# is 118.94 mm^2, and the evaluation result is 116.13 mm^2, with an error of 2.4%, achieving high reconstruction accuracy. The above shows that the segmentation interpolation algorithm based on the characteristic signal Bx can achieve high-precision reconstruction and visual display of the crack profile.

6 Conclusion

This paper establishes a three-dimensional finite element simulation model of ACFM in a seawater environment, analyzes in detail the influence of crack length and depth on characteristic signals, proposes a two-step interpolation crack size evaluation algorithm and a segmentation interpolation crack profile reconstruction algorithm, builds an underwater ACFM testing system, and carries out crack detection and evaluation experiments. The main conclusions are as follows:

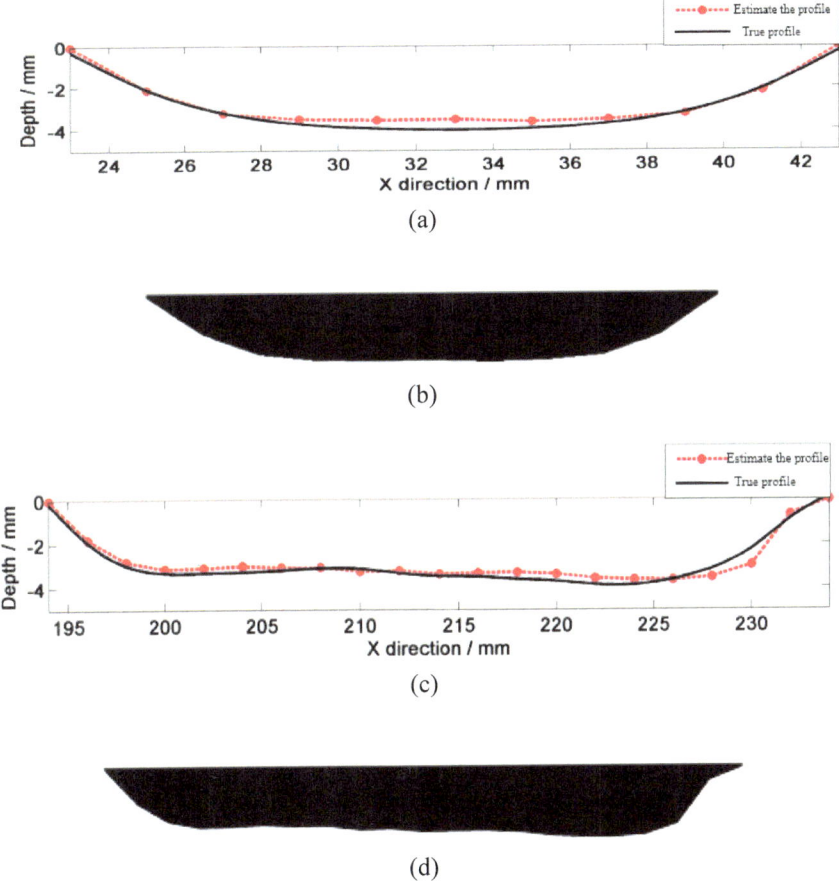

Fig. 10 Reconstruction results of crack profile. **a** 1# Crack reconstruction results. **b** 1# Crack profile visualization morphology. **c** 2# Crack reconstruction results. **d** 2# Crack profile visualization morphology

(1) The distance between the peaks and valleys of the characteristic signal Bz is related to the crack length and is not affected by the crack depth; the depth of the characteristic signal Bx valley is mainly related to the crack depth, but is also affected by the crack length.

(2) When the crack is larger than 30 mm, the crack length does not affect the depth of the characteristic signal Bx valley; when the crack is less than 30 mm, the crack length has a greater impact on the characteristic signal Bx valley.

(3) The two-step interpolation crack size evaluation algorithm based on the characteristic signals Bx and Bz can achieve high-precision evaluation of crack length and depth, with a maximum error of 3.0% for length evaluation and a maximum error of 7.5% for depth evaluation.

(4) The segmentation interpolation profile reconstruction algorithm based on the characteristic signal Bx can achieve high-precision reconstruction and visual display of the crack profile, with a maximum reconstruction error of 5.7%.

References

1. M.P. Papaelias, M.C. Lugg, C. Roberts et al., High-speed inspection of rails using ACFM techniques. NDTE Int. **42**, 328–335 (2009)
2. W. Li, X. Yuan, G. Chen et al., High sensitivity rotating alternating current field measurement for arbitrary-angle underwater cracks. NDTE Int. **79**, 123–131 (2016)
3. J. Leng, H. Tian, G. Zhou et al., Joint detection of MMM and ACFM on critical parts of jack-up offshore platform. Ocean Eng. **35**, 34–38 (2017)
4. A. Akbari-Khezri, S.H.H. Sadeghi et al., An efficient modeling technique for analysis of AC field measurement probe output signals to improve crack detection and sizing in cylindrical metallic structures. J. Nondestruct. Evaluat. **35**, 9 (2016)
5. J.L. Shen, L. Zhou, H. Rowshandel, G.L. Nicholson, C.L. Davis, Determining the propagation angle for non-vertical surface-breaking cracks and its effect on crack sizing using an ACFM sensor. Measur. Sci. Technol. **26**, 115604 (2015)
6. Z. Zhou, J. Zhang, C. Gu, Simulation research on crack defect identification based on alternating current electromagnetic field. J. Saf. Sci. Technol. **14**(12), 146–151 (2018)
7. X. Yuan, W. Li, G. Chen et al., Two-step interpolation algorithm for measurement of longitudinal cracks on pipe strings using circumferential current field testing system. IEEE Trans. Ind. Inform. **14**, 394–402 (2017)
8. A.M. Lewis, D.H. Michael, M.C. Lugg et al., Thin-skin electromagnetic fields around surface-breaking cracks in metals. J. Appl. Phys. **64**(8), 3777–3784 (1988)
9. W.D. Dover, R. Collins, D.H. Michael et al., The use of AC-field measurements for crack detection and sizing in air and underwater [and discussion]. Philos. Trans. R. Soc. Lond. A Math. Phys. Sci. **320**(1554), 271–283 (1986)
10. S. Hu, *Research on Detection and Inversion of Cracks Using the Alternating Current Field Measurement Technique* (Tianjin University, Tianjin, 2004)
11. A.L. Ribeiro, H.G. Ramos, O. Postolache, A simple forward direct problem solver for eddy current non-destructive inspection of aluminum plates using uniform field probes. Measurement **45**, 213–217 (2012)
12. H. Rowshandel, G.L. Nicholson, J.L. Shen et al., Characterisation of clustered cracks using an ACFM sensor and application of an artificial neural network. NDTE Int. **98**, 80–88 (2018)
13. G.L. Nicholson, C.L. Davis, Modelling of the response of an ACFM sensor to rail and rail wheel RCF cracks. NDTE Int. **46**, 107–114 (2012)
14. R.K. Amineh, M. Ravan, S.H.H. Sadeghi et al., Using AC field measurement data at an arbitrary liftoff distance to size long surface-breaking cracks in ferrous metals. NDTE Int. **41**, 169–177 (2008)
15. D. Katoozian, R.P.R. Hasanzadeh, A fuzzy error characterization approach for crack depth profile estimation in metallic structures through ACFM data. IEEE Trans. Magnet. **53**, 6202110 (2017)
16. W. Li, G. Chen, X. Zheng, Crack sizing for alternating current field measurement based on GRNN. J. China Univ. Petrol. **31**(2), 105–109 (2007)
17. W. Li, X. Yuan, M. Qu et al., Research on real-time and high-precision cracks inversion algorithm for ACFM based on GA-BP neural network. J. China Univ. Petrol. **40**(5), 128–134 (2016)
18. A. Noroozi, R.P.R. Hasanzadeh, M. Ravan, A fuzzy learning approach for identification of arbitrary crack profiles using ACFM technique. IEEE Trans. Magnet. **49**(9), 5016–5027 (2013)

19. X. Yuan, W. Li, G.M. Chen et al., Inspection of both inner and outer cracks in aluminum tubes using double frequency circumferential current field testing method. Mech. Syst. Signal Process. **127**, 16–34 (2019)
20. G.L. Nicholson, A.G. Kostryzhev, X.J. Hao et al., Modelling and experimental measurements of idealised and light-moderate RCF cracks in rails using an ACFM sensor. NDTE Int. **44**, 427–437 (2011)
21. X. Yuan, W. Li, G.M. Chen et al., Circumferential current field testing system with TMR sensor array for non-contact detection and estimation of cracks on power plant piping. Sens. Actuat. A **263**, 542–553 (2017)
22. X. Yuan, W. Li, G.M. Chen et al., Bobbin coil probe with sensor arrays for imaging and evaluation of longitudinal cracks inside aluminum tubes. IEEE Sens. J. **18**, 6774–6781 (2018)